Stem Cell Laboratory Techniques

Stem Cell Laboratory Techniques

A Guide for Researchers and Students

Edited by

Suresh Kumar Subbiah

Centre for Materials Engineering and Regenerative Medicine, Bharath Institute of Higher Education and Research, Chennai, Tamil Nadu, India

Akon Higuchi

State Key Laboratory of Ophthalmology, Optometry and Visual Science, Eye Hospital, Wenzhou Medical University, Wenzhou, Zhejiang, China

Department of Chemical and Materials Engineering, National Central University, Taoyuan, Taiwan

Center for Emergent Matter Science, Riken, Wako, Saitama, Japan

Pooi Ling Mok

Department of Biomedical Science, Faculty of Medicine and Health Sciences, Universiti Putra Malaysia, Serdang, Selangor, Malaysia

ELSEVIER

ACADEMIC PRESS

An imprint of Elsevier

Academic Press is an imprint of Elsevier
125 London Wall, London EC2Y 5AS, United Kingdom
525 B Street, Suite 1650, San Diego, CA 92101, United States
50 Hampshire Street, 5th Floor, Cambridge, MA 02139, United States
The Boulevard, Langford Lane, Kidlington, Oxford OX5 1GB, United Kingdom

ISBN 978-0-12-823729-8

For information on all Academic Press publications
visit our website at https://www.elsevier.com/books-and-journals

Publisher: Stacy Masucci
Acquisitions Editor: Elizabeth Brown
Editorial Project Manager: Tracy Tufaga
Production Project Manager: Punithavathy Govindaradjane
Cover Designer: Matthew Limbert

Typeset by STRAIVE, India

Contents

Contributors

Khairul Akmal Abdul Rahman
Department of Biomedical Science, Faculty of Medicine and Health Sciences, Universiti Putra Malaysia, Serdang, Selangor, Malaysia

Aida Suraya Binti Abu Yazid
Faculty of Science, University of Malaya, Kuala Lumpur, Malaysia

Mathanakumara Ealam Selvan
Faculty of Science, University of Malaya, Kuala Lumpur, Malaysia

Nurfarhana Ferdaos
Department of Pharmacology and Chemistry, Faculty of Pharmacy, Universiti Teknologi MARA Puncak Alam Campus, Selangor, Malaysia

Adila A. Hamid
Department of Physiology, Faculty of Medicine, National University of Malaysia Medical Centre, Kuala Lumpur, Malaysia

Akon Higuchi
State Key Laboratory of Ophthalmology, Optometry and Visual Science, Eye Hospital, Wenzhou Medical University, Wenzhou, Zhejiang, China; Department of Chemical and Materials Engineering, National Central University, Taoyuan, Taiwan; Center for Emergent Matter Science, Riken, Wako, Saitama, Japan

Tong Jiabei
Department of Biomedical Science, Faculty of Medicine and Health Sciences, Universiti Putra Malaysia, Serdang, Selangor, Malaysia

Muhammad Khair Joharry
Department of Biomedical Science, Faculty of Medicine and Health Sciences, Universiti Putra Malaysia, Serdang, Selangor, Malaysia

Thilakavathy Karuppiah
Department of Biomedical Science; Genetics & Regenerative Medicine (ReGEN) Research Group, Faculty of Medicine and Health Sciences, Universiti Putra Malaysia, Serdang, Selangor, Malaysia

Avin Ee-Hwan Koh
Department of Biomedical Science, Faculty of Medicine and Health Sciences, Universiti Putra Malaysia, Serdang, Selangor, Malaysia

Nur Izzati Mansor
Department of Biomedical Science, Faculty of Medicine and Health Sciences, Universiti Putra Malaysia, Serdang, Selangor, Malaysia

John O. Mason
Simons Initiative for the Developing Brain and Centre for Discovery Brain Sciences, University of Edinburgh, Edinburgh, United Kingdom

Devasharmini Mathandaver
Department of Biomedical Science, Faculty of Medicine and Health Sciences, Universiti Putra Malaysia, Serdang, Selangor, Malaysia

Pooi Ling Mok
Department of Biomedical Science, Faculty of Medicine and Health Sciences, Universiti Putra Malaysia, Serdang, Selangor, Malaysia

Siti Sarah Mustaffa Al Bakri
Department of Biomedical Science, Faculty of Medicine and Health Sciences, Universiti Putra Malaysia, Serdang, Selangor, Malaysia

Norshariza Nordin
Department of Biomedical Science; Genetics & Regenerative Medicine (ReGEN) Research Group, Faculty of Medicine and Health Sciences, Universiti Putra Malaysia, Serdang, Selangor, Malaysia

Ghee Chien Ooi
CryoCord Sdn. Bhd., Cyberjaya, Selangor, Malaysia

Hernandez Maradiaga Oscar David
Department of Medical Microbiology, Faculty of Medicine and Health Sciences, Universiti Putra Malaysia, Serdang, Selangor, Malaysia

Zulida Rejali
Department of Obstetrics and Gynaecology, Faculty of Medicine and Health Sciences, Universiti Putra Malaysia, Serdang, Selangor, Malaysia

Sakinah-Syed
Department of Medical Microbiology, Faculty of Medicine and Health Sciences, Universiti Putra Malaysia, Serdang, Selangor, Malaysia

Suresh Kumar Subbiah
Centre for Materials Engineering and Regenerative Medicine, Bharath Institute of Higher Education and Research, Chennai, Tamil Nadu, India

Baskar Subramani
Medixcell Sdn Bhd, Bandar Sg Long, Kajang, Malaysia

Seoh Wei Teh
Department of Orthopaedics, Faculty of Medicine and Health Sciences, Universiti Putra Malaysia, Serdang, Selangor, Malaysia

Mohd Nazri Yazid
Department of Obstetrics and Gynaecology, Faculty of Medicine and Health Sciences, Universiti Putra Malaysia, Serdang, Selangor, Malaysia

Sun Zhong
Genetics & Regenerative Medicine (ReGEN) Research Group, Faculty of Medicine and Health Sciences, Universiti Putra Malaysia, Serdang, Selangor, Malaysia

About the editors

Suresh Kumar Subbiah

Pro Vice-Chancellor, Centre for Materials Engineering and Regenerative Medicine, Bharath Institute of Higher Education and Research, Chennai, Tamil Nadu, India

Suresh Kumar Subbiah previously worked as a Professor at Universiti Putra Malaysia (UPM), Malaysia. He was a Postdoctoral Research Fellow at the National Taiwan University, Taiwan and National Central University, Taiwan in the fields of yeast genetics and stem cells. He has also worked as a Senior Research Executive (Fermentation of Microbial Drugs) in IPCA Laboratory Ltd., Mumbai, India; an R&D officer (Fermentation of Microbial Drugs) at Gujarat Themis Biosyn Ltd., Gujarat, India; and a Senior and Junior Research Fellow (Fermentation and Purification of Enzymes) at the University of Delhi South Campus, New Delhi, India. Currently, he is an Academic Editor of PLOS ONE journal, an academic editor in Current pharmaceutical biotechnology, a Special Issue Editor in Frontiers in Pharmacology, an Associate Editor in Progress in Bioscience and Bioengineering, and an Editor of the American Journal of Tissue Engineering, Columbia International Publishing. His research interests include multidisciplinary areas such as stem cells for treating infectious diseases, cancer stem cells, stem cell niches, induced pluripotent stem cells, purification of human induced pluripotent stem cells, and biomaterials for stem cell culture and differentiation. He has more than 250 publications in peer-reviewed journals and has the highest impact factor of 48.5, with h-index of 40 and i10-index of 120.

Akon Higuchi

State Key Laboratory of Ophthalmology, Optometry and Visual Science, Eye Hospital, Wenzhou Medical University, Wenzhou, Zhejiang, China
Department of Chemical and Materials Engineering, National Central University, Taoyuan, Taiwan
Center for Emergent Matter Science, Riken, Wako, Saitama, Japan

Akon Higuchi is a Chair Professor at National Central University (Taiwan). He is also joined as a Visiting Professor at Eye Hospital, Wenzhou Medical University. He received his Ph.D. from the Tokyo Institute of Technology. He was a

Professor at Seikei University (Tokyo). He is a fellow of Royal Society of Chemistry (England) and American Institute for Medical & Biological Engineering (AIMBE, United States). He is on the editorial board of laboratory investigation (Nature Springer), Tissue Engineering and Regenerative Medicine and Journal of Materials Chemistry B, Editor of Open Physics, Associate Editor of IET Nanobiotechnology, and TERMIS-AP council. He is interested in the development of biomaterials for stem cell research. He received an Outstanding Scientific Award from TERMIS-AP. He established a purification method of hematopoietic stem cells and mesenchymal stem cells from umbilical cord blood and adipose tissue, respectively, by filtration method through polymeric membranes. He is also establishing universal human induced pluripotent stem cells from fetal cells. He is also developing biomaterials for stem cell culture and differentiation. He published over 300 SCI articles and 3 articles in Chemical Reviews as first and corresponding author.

Pooi Ling Mok
Department of Biomedical Science, Faculty of Medicine and Health Sciences, Universiti Putra Malaysia, Serdang, Selangor, Malaysia

Pooi Ling Mok received her Ph.D. award from the National University of Malaysia, Malaysia in 2012. She joined the Faculty of Medicine and Health Sciences, Universiti Putra Malaysia, Malaysia in 2013 and is currently an Associate Professor in the Department of Biomedical Science. She also previously served as an Assistant Professor at Jouf University, Saudi Arabia, from 2017 to 2021. She is also the founding Vice President of the Malaysian Stem Cell for Research and Therapy (MSCRT) association. Dr. Mok has a passion for research involving stem cell engineering at the genetic, cell, and tissue levels involving human samples with the aim of improving the quality of life in patients. Most of her research works integrated many methods that include genetic modifications, the use of biomaterials, and tissue engineering for optimal result outcomes. Her current research focuses include finding alternative treatments for visual impairment, delaying aging, cardiovascular and autoimmune diseases, and cancers using stem cell-based products and immune cells.

CHAPTER

1

Introduction: Stem cells and their application in research and therapy

Tong Jiabei[a], Pooi Ling Mok[a], and Suresh Kumar Subbiah[b]

[a]*Department of Biomedical Science, Faculty of Medicine and Health Sciences, Universiti Putra Malaysia, Serdang, Selangor, Malaysia,* [b]*Centre for Materials Engineering and Regenerative Medicine, Bharath Institute of Higher Education and Research, Chennai, Tamil Nadu, India*

1. Definition and characteristics of stem cells

Stem cells are a group of cells that possess the unique characteristics of undergoing self-renewal and differentiation. They can proliferate in vitro and also differentiate into specific kinds of cells. During the development of the oosperm, the majority of the cells will undergo differentiation and maturation to form cells with specific phenotypes and functions. A small number of undifferentiated cells will be preserved in various tissues as a reservoir for tissue maintenance. The microenvironment in which stem cells are present is known as "stem cell niche." This niche consists of signals that could modulate the proliferation and differentiation of stem cells. When there is a loss of tissue due to physical or chemical injury, for example, the stem cells will be activated to divide and differentiate into specific cells to replace the damaged tissue (Bacakova et al., 2018; Kolagar et al., 2020; Sameri et al., 2020).

In the area of regenerative medicine, stem cells can be used as an important strategy for cell or tissue replacement in the treatment of various conditions such as traumatism, cardiovascular disease, neurodegenerative diseases, endocrine disease, and other diseases. Additionally, human stem cells can also be used as research models to explore the various growth and differentiation mechanisms of cells and to study the processes involved in human development, growth, and diseases (Bourgeois et al., 2021; Elder and Dale, 2019; Fagan, 2019; Zakrzewski et al., 2019). Therefore, stem cell-related research has significant application value in cytotherapy, organ transplantation, and drug screening.

Stem Cell Laboratory Techniques. https://doi.org/10.1016/B978-0-12-823729-8.00008-7

2. Classification of stem cells

2.1 Classification based on differentiation potential

Oosperm cleavage produces morula with 16–32 cells and then continues to develop into a blastocyst. The blastocyst can differentiate into the endoderm, mesoderm, and ectoderm. The cells derived from the different germ layers will differentiate into tissue specific progenitor/stem cells or form diverse adult cells. As tissues are injured by trauma, senility, diseases, or other factors, these progenitor/stem cells can play a role in facilitating the repairing process in the injured tissues. Based on the differentiation potential, stem cells can be classified into the following three types (Fig. 1):

(1) Totipotent stem cells: These are stem cells that can efficiently differentiate and develop into a complete individual. Totipotent stem cells can be generated from the zygote, which can divide into more cells and generate blastocyst. The human cells in the 16-cell stage after zygote division is totipotent (Baker and Pera, 2018; Cai et al., 2022).

(2) Pluripotent stem cells: Unlike totipotent stem cells, pluripotent stem cells have restricted potency to differentiate and develop into a complete individual. However, these stem cells can differentiate and form various cell tissues. For example, embryonic stem cells (ESCs) can proliferate infinitely and

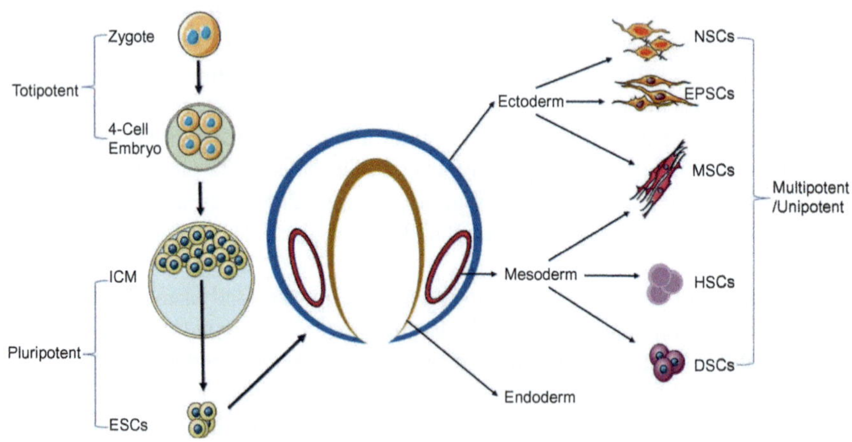

FIG. 1

Stem cell: Classification based on differentiation potential. Totipotent stem cells include zygotes and the cells of the 4-cell stage embryo, which can develop into the three primary germ cell layers and into extra-embryonic tissues such as the placenta. Differentiation begins at the 8-cell stage embryo. ESCs are the pluripotent stem cells derived from the blastocyst ICM, which can differentiate into any of the three germ layers. *DSCs*, dermis-derived stem cells; *EPSCs*, epidermal stem cells; *ESCs*, embryonic stem cells; *ICM*, inner cell mass (in blastocyst); *MSCs*, mesenchymal stem cells; *NSCs*, neural stem cells.

differentiate into over 200 types of cells in different tissues and organs, and then further aid in the formation of all the tissues and organs of an organism (Le et al., 2020; Riveiro and Brickman, 2020; Yamanaka, 2020). Other than that, mesenchymal stem cells (MSCs) in the bone marrow can differentiate into adipogenic, osteogenic, and chondrogenic cell lineages.

(3) Unipotent stem cells: Also known as committed stem cells, the unipotent stem cells can only differentiate into one or two closely related types of somatic cells, for example, skin fibroblast, vascular endothelial progenitor cells, and osteoblast. Most precursor cells belong to this class of stem cells (Bahmad et al., 2021; Sundaravadivelu et al., 2021; Xu et al., 2019).

2.2 Classification based on cell sources

Based on the developmental stage of embryonic development, stem cells can be further divided into ESCs and adult stem cells (ASCs). The somatic cells can also be reprogrammed to achieve pluripotency by forced expression of specific transcription factors, direct nuclear transfer from a somatic cells into an enucleated egg, or somatic cell fusion with pluripotent cells (Laplane and Solary, 2019; Zakrzewski et al., 2019).

ESCs are the cells derived from the inner mass of the blastocyst, which predominantly exists in the undifferentiated state. These cells have developmental potency to differentiate into most types of cells in adult animals, including germ cells. Besides, embryonic germ stem cells, trophoblast stem cells, and some other stem cells can also be isolated from other parts intra and extra of an embryo (Suman et al., 2019; Yamanaka, 2020).

Adult stem cells exist in the tissues and organs of adult animals, such as the nerve, skin, and marrow, and possess the potency of repairing as well as regenerating. In an adult human body, ASCs primarily exist in a specific microenvironment and can maintain a state of dormacy for a long time without entering the cell differentiation cycle. These cells can also generate new stem cells or differentiate by using certain procedures and form new cells with a specific function to maintain the functional state of tissues and organs (Clevers and Watt, 2018; Gurusamy et al., 2018).

2.3 The classical types of stem cells

2.3.1 Embryonic stem cells (ESCs)

After the ovum is fertilized and an oosperm is formed, oosperm goes through a series of cleavage reactions to produce a set of cells called blastomeres. These cells further can effectively divide and rearrange to form a blastocyst, which consists of an inner layer of cells called the embryonic knob and an outer layer of cells called the trophoblast. The cells in the inner cell mass can eventually develop into the different tissues and organs of the body, so it can be an ideal source of ESCs (Kaufman and Thomson, 2019; Muhammad, 2020; Yamanaka, 2020). But isolation of ESCs from inner cell mass will destroy embryos. Since opponents of embryo research consider that deriving the stem cells destroys the blastocyst, an unimplanted human embryo at the sixth to the eighth day of development, this made the application of inner cell mass face ethical issues. So the researchers prefer to obtaining these cells from a single blastomere biopsy (Klimanskaya et al., 2006).

ESCs can form dense spherical cell colonies in the culture medium and express several transcription factors and cell surface markers, which include stage-specific teratocarcinoma-derived growth factor 1 (TDGF1), stage specific embryonic antigen 3/4 (SSEA-3/4), octamer-binding transcription factor 3/4 (Oct-3/4), and Nanog (Mora et al., 2017; Suman et al., 2019).

2.3.2 Mesenchymal stem cells (MSCs)

Mesenchymal stem cells (MSCs) are stem cells that primarily exist in mesenchyma of adult tissues and retain the potency of proliferation and differentiation, including bone marrow mesenchymal stem cells (BMSCs), adipose tissue-derived stromal cells (ADSCs), umbilical cord mesenchymal stem cells (UMSCs), and dental pulp stem cells (DPSCs). MSCs can display the dual functions of self-renewal and differentiation, and they can be differentiated into many kinds of adult cells, which are widely used in stem cell therapy. The dynamic balance of the two abilities can aid to maintain the normal fate of the organization. Otherwise, excessive differentiation and insufficient self-renewal can lead to stem cell depletion. Moreover, excessive self-renewal will hinder differentiation, stem cell expansion, and even lead to formation of teratoma. The prevention of excessive differentiation and insufficient self-renewal depend on the gene expression regulation of stem cells (Liu et al., 2020b; Saeedi et al., 2019; Zhang et al., 2021).

However, compared with other kinds of stem cells, MSCs have the following application advantages: (1) there are no limitations related to ethical applications. Compared with embryonic stem cells, which have more comprehensive cell differentiation potential, the use of adult stem cells is not subject to ethical restrictions as the acquisition of cells does not involve the destruction of embryo. (2) they display greater ability of directional differentiation. (3) the majority of MSCs do not exhibit immunogenicity, so it can avoid immune rejection in allograft transplantation (Brown et al., 2019; Liu et al., 2020b; Samadi et al., 2021; Zhang et al., 2021).

2.3.3 Induced pluripotent stem cells (iPSCs)

Induced pluripotent stem cells (iPSCs) are pluripotent cells derived from the reprogramming of the differentiated somatic cells (e.g., skin fibroblasts, peripheral blood monocytes, and epithelial cells in urine) and can display similar pluripotency to ESCs. In 2006, Takahashi and Yamanaka found that the introduction of four specific genes (*Oct4*, *c-Myc*, *Sox2*, and *Klf4*) into mice fibroblast cells enabled these cells to possess the specific features of ESCs. The different methods of reprogramming include the following: (1) application of the retroviruses to make the cells in an intermediate reprogramming state, (2) plasmid transfection, and (3) induction by small RNA or small molecular protein (Carpenedo et al., 2016; Rajan, 2014). Currently, iPSCs have been obtained from the somatic cells of many different patients. These iPSCs and differentiation products can be used for cell replacement treatments or screening of potential therapeutic drugs. Despite that, the molecular mechanisms of cell reprogramming to iPSC are still not fully understood. More effort has been focused on improving the reprogramming efficiency across many different cell types. The traditional gene manipulation, delivery of molecule proteins could be more convenient and safe delivery of to generate iPSCs for clinical usage (Doss and Sachinidis, 2019; Mobley, 2019; Yamanaka, 2020; Zaehres, 2020).

3. Research and application of stem cells

By applying different inducer agents in a culture for a period of time, stem cells can be stimulated to form different type of cells such as neuron cells, hepatocytes, endothelial, myocardial, islet β or hematopoietic cells, etc. Hence, the stem cells and their differentiated products can be of important application value in the following fields:

3.1 Developmental biology research

Stem cells can be primarily used as a research model to reveal the mechanisms of development in humans and animals. Many developmental events cannot be studied in the human embryos. Embryonic stem cells have the potency of multidirectional differentiation and possess the capacity to form embryoid bodies, so they can be used to recapitulate the developmental process of early embryos and facilitate the study of human embryo development, cell differentiation, function establishment, aging, death, and other important physiological processes. iPSCs have been functionally used to elucidate the role of specific genes, growth factors, and cytokines, as well as to study the gene expression of certain cells or tissues during the development process such as neurocyte and germ cells (Bratt-leal et al., 2020; Martinez-Gamero et al., 2021).

3.2 Application in drug screening and development

Stem cells can be obtained in vitro by expanding the culture of different cells or can induce them to differentiate into specific cells, which can be used as experimental models for drug screening in vitro. This approach can reduce the overall cost associated with the drug development process and overcome potential ethical challenges in animal models. At the same time, human cells can be effectively used to screen drugs, and the species variations between animals and humans can increase the safety and reliability of drugs. The embryonic stem cell test (EST), officially approved by the European Centre for Alternative Animal Experiments (ECVAM) as an in vitro alternative method for drug and compound toxicity screening, has entered the stage of practical application and has become an internationally recognized in vitro alternative method for analysis of cell line developmental toxicity (Seiler and Spielmann, 2011).

3.3 Application in regenerative medicine

As damaged tissue or organs cannot self-heal, stem cell transplantation and tissue engineering are ideal choices to overcome these limitations. Stem cell transplantation has been extensively used in clinical medicine, such as hematopoietic stem cell transplantation for leukemia. In addition, several new form of treatment modalities using stem cells are currently undergoing clinical trials, such as transplantation of MSC-induced insulin-producing cell transplantation for diabetes, differentiated nerve cells replacement to ameliorate neurodegeneration in Alzheimer's disease, Parkinson's disease, Huntington's disease, etc., or direct injection of MSCs for nephropathy. It has been found that in transplantation medicine, the immune rejection of allograft has always been a significant barrier. Stem cell transplantation can overcome this major obstacle of immune rejection in two distinct ways: (1) the immunogenicity of some stem cells is weak, thereby leading to reduced rejection severity; (2) autologous stem cells can be easily isolated and expanded in vitro before retransplantation into the patient; (3) somatic cells from an adult patient can be isolated and reprogrammed to form iPSCs, and thereafter redifferentiated into required cells before transplantation into the patient (George et al., 2018; Jeeva Bharathi et al., 2019).

Tissue engineering employs biochemical as well as physicochemical factors to reconstruct cell tissues using cells and biomaterial scaffolds, under to replace the lost biological tissues in patients. The tissue cells derived from some part of the human body can be cultured in vitro and expanded to high concentration, and then adsorbed on a biological material with ideal biocompatibility and degradation rate as well as absorbability. Stem cells are an ideal population to be used as seed cells for tissue engineering due to their multidirectional differentiation potential and the ability to expand under in vitro settings (Bakhshandeh et al., 2017; Pearson et al., 2017; Ramos and Moroni, 2020) (Table 1).

Table 1 Stem cells from different sources and their characteristics.

Type of stem cells	Source	Cell surface marker/gene	Characteristic
Embryonic stem cells (ESCs) (Cai et al., 2022)	Inner cell mass of embryo	Oct-4, Nanog, Sox2, fgf-4	Unlimited capacity of self-renewal; pluripotential
Induced pluripotent stem cells (iPSCs) (Liu et al., 2020a)	Reprogramed from Dermal fibroblast, Myoblast, epithelial cell et al.	Oct-4, Nanog, Sox2	Unlimited capacity for self-renewal; pluripotential
Hematopoietic stem cells (HSCs) (Poulos, 2018)	Bone marrow	$CD34^+$, $EMCN^+$, $CD34^+$, $CD59^+$, Thy1/$CD90^+$, C-kit/$CD117^+$	High proliferative capacity; pluripotency to differentiate to all types of blood cells

Table 1 Stem cells from different sources and their characteristics—cont'd

Type of stem cells	Source	Cell surface marker/gene	Characteristic
Umbilical cord blood MSCs (UCB-MSCs) (Um, 2020)	Umbilical cord blood	$CD10^+$, $CD13^+$, $CD29^+$, $CD44^+$, $CD73^+$, $CD90^+$, $CD105^+$, CD11B/$CD24^-$, $CD34^-$, $CD45^-$, HLA-DR^-	High proliferative capacity; pluripotency, wide source, low immunogenicity
Bone marrow mesenchymal stem cells (BMSCs) (Chu, 2020)	Bone marrow	$CD106^+$, $CD73^+$, $CD90^+$, $CD105^+$, CD11B/$CD24^-$, $CD34^-$, $CD45^-$, HLA-DR^-	High proliferative capacity; pluripotency
Adipose-derived stem cells (ADSCs) (Bacakova et al., 2018)	Adipose tissue	$CD49d^+$, $CD73^+$, $CD90^+$, $CD105^+$, CD11B/$CD24^-$, $CD34^-$, $CD45^-$, HLA-DR^-	High proliferative capacity; pluripotency, wide source
Dental pulp stem cells (DPSCs) (Tsutsui, 2020)	Dental pulp tissue	$CD73^+$, $CD90^+$, $CD105^+$, CD11B/$CD24^-$, $CD34^-$, $CD45^-$, HLA-DR^-	High proliferative capacity; pluripotency potential lower that other MSCs
Epidermal stem cells (Lenkiewicz, 2019)	Epidermal tissue	Integrin $\alpha6^+$, Integrin $\beta1^+$, $10G7^+$, $P63^+$	High capacity for self-renewal; slow cycling, adhere to basement membrane
Dermal stem cells (Swamy, 2021)	Dermal tissue	Integrin $\alpha6^+$, $CD49f^+$, $CD34^+$	High proliferative capacity; pluripotency for skin cells; neuronal differentiation ability
Endothelial progenitor cells (EPCs)/human umbilical vein endothelial cells (HUVECs) (Peters, 2018)	Vascular endothelium/ peripheral blood/ umbilical cord blood/umbilical cord vessel	$CD14^+$, VEGFR-3^+, $CD31^+$, $CD34^+$, $CD144^+$, Von $Willebrand^+$	Vascular endothelial cell differentiation ability
Corneal stem cells/ limbal stem cells (Nurković, 2020)	Limbal tissue	$ABCG2^+$, $P63^+$, Integrin $\alpha9^+$, $Keratin3^-$, $Keratin12^-$	Corneal epithelial cell differentiation ability

Continued

Table 1 Stem cells from different sources and their characteristics—cont'd

Type of stem cells	Source	Cell surface marker/gene	Characteristic
Neural stem cells (NSCs) (Grochowski, 2018)	Nervous tissue	MCM2$^+$, PCNA$^+$, BLBP$^+$, GFAP$^+$, Sox2$^+$, Nestin$^+$	Differentiate to neuronal restricted precursor/glial restricted precursor
Endogenous lung stem/progenitor cells (McQualter, 2019)	Weasand/ Bronchus/ Pulmonary alveoli	Basal cell—K5$^+$, K14$^+$, P63$^+$ Duct cell—k5$^+$, k14$^+$ Clara cells—CCSP$^+$, CYP450 2F2$^+$ BASC—CCSP$^+$, PRO-SPC$^+$	Differentiate to weasand/ bronchus/ pulmonary alveoli cells
Cardiac stem cell (Guo and Pu, 2020)	Heart	c-Kit/Sca-1/Isl-1$^+$	Myocardium repair
Spermatogonial stem cells (SSCs) (Kubota, 2018)	Male genital	Integrin α6$^+$, Integrin β1$^+$, c-Kit$^+$, AKP$^+$, SSEA-1$^+$	High pluripotency after dedifferentiation
Hepatic stem cells (Chen, 2018)	Liver tissue-Canal of Hering	CK19$^+$, NVAM$^+$, Ep-CAM$^+$, CLDN-3$^+$, CD90$^+$	Differentiate to all types of liver cells
Intestinal stem cells (ISCs) (Zhao, 2020)	Intestinal crypt	Lgr4/5$^+$	Differentiate to enterocyte

4. Conclusion

Stem cells have shown their remarkable advantages in the areas of tissue repairing, developmental biology research, disease modeling, and drug screening; however, these cells also presented some challenges and drawbacks. The clinical application of ESCs is severely limited due to the higher tumorigenicity, ethical restrictions, and hard-to-maintain undifferentiated status. The development of iPSCs has resolved ethical issues and the barriers of immunological rejection in a transplantation, providing a new source of cells for clinical research and application. However, the efficiency iPSCs generated from different somatic cells or tissues in different developmental stages varies vastly, thereby warranting the need to validate the safety iPSCs for future clinical application. Research on stem cells still needs improvement in the following areas: (1) elucidating the intrinsic mechanisms of cell proliferation and differentiation to facilitate the establishment and application of disease and drug screening models, (2) developing mature gene editing techniques to improve the stability and safety of iPSCs technology, and (3) improving the standardization in the production for better stem cell reproducibility and its application.

Abbreviations

ADSCs	adipose tissue-derived stromal cells
ASCs	adult stem cells
BMSCs	bone marrow mesenchymal stem cells
DPSCs	dental pulp stem cells
ECVAM	European Centre for Alternative Animal Experiments
EPCs	endothelial progenitor cells
ESCs	embryonic stem cells
EST	embryonic stem cell test
HSCs	hematopoietic stem cells
HUVECs	human umbilical vein endothelial cells
iPSCs	induced pluripotent stem cells
MSCs	mesenchymal stem cells
Oct-3/4	octamer-binding transcription factor 3/4
SSEA-3/4	stage specific embryonic antigen 3/4
TDGF1	teratocarcinoma-derived growth factor 1
UMSCs	umbilical cord mesenchymal stem cells

References

Bacakova, L., Zarubova, J., Travnickova, M., Musilkova, J., Pajorova, J., Slepicka, P., et al., 2018. Stem cells: their source, potency and use in regenerative therapies with focus on adipose-derived stem cells—a review. Biotechnol. Adv. 36, 1111–1126. https://doi.org/10.1016/j.biotechadv.2018.03.011.

Bahmad, H.F., Elajami, M.K., Daouk, R., Jalloul, H., Darwish, B., Chalhoub, R.M., et al., 2021. Stem cells: in sickness and in health. Curr. Stem Cell Res. Ther. 16, 262–276. https://doi.org/10.2174/1574888X15999200831160710.

Baker, C.L., Pera, M.F., 2018. Capturing totipotent stem cells. Cell Stem Cell 22 (1), 25–34. https://doi.org/10.1016/j.stem.2017.12.011.

Bakhshandeh, B., Zarrintaj, P., Oftadeh, M.O., Keramati, F., Fouladiha, H., Sohrabi-Jahromi, S., et al., 2017. Tissue engineering; strategies, tissues, and biomaterials. Biotechnol. Genet. Eng. Rev. 33, 144–172. https://doi.org/10.1080/02648725.2018.1430464.

Bourgeois, S., Sawatani, T., Van Mulders, A., De Leu, N., Heremans, Y., Heimberg, H., et al., 2021. Towards a functional cure for diabetes using stem cell-derived beta cells: are we there yet? Cell 10. https://doi.org/10.3390/cells10010191.

Bratt-leal, A.M., Zhang, A., Wang, Y., Loring, J.F., 2020. Applications for stem cells. In: Principles of Tissue Engineering. Elsevier, Inc, pp. 445–456, https://doi.org/10.1016/B978-0-12-818422-6.00025-3.

Brown, C., McKee, C., Bakshi, S., Walker, K., Hakman, E., Halassy, S., et al., 2019. Mesenchymal stem cells: cell therapy and regeneration potential. J. Tissue Eng. Regen. Med. 13, 1738–1755. https://doi.org/10.1002/term.2914.

Cai, J., Chen, H., Xie, S., Hu, Z., Bai, Y., 2022. Research progress of totipotent stem cells. Stem Cells Dev. 31 (13–14), 335–345. https://doi.org/10.1089/scd.2022.0061.

Carpenedo, R., Julian, L., Stanford, W., 2016. Human pluripotent stem cells. Methods 101. https://doi.org/10.1016/j.ymeth.2016.04.012.

Chen, C., et al., 2018. Biotechnology challenges to in vitro maturation of hepatic stem cells. Gastroenterology 154 (5), 1258–1272. https://doi.org/10.1053/j.gastro.2018.01.066.

Chu, D.-T., 2020. An update on the progress of isolation, culture, storage, and clinical application of human bone marrow mesenchymal stem/stromal cells. Int. J. Mol. Sci. 21 (3), 708. https://doi.org/10.3390/ijms21030708.

Clevers, H., Watt, F.M., 2018. Defining adult stem cells by function, not by phenotype. Annu. Rev. Biochem. 87, 1015–1027. https://doi.org/10.1146/annurev-biochem-062917-012341.

Doss, M.X., Sachinidis, A., 2019. Current challenges of iPSC-based disease modeling and therapeutic implications. Cell 8. https://doi.org/10.3390/cells8050403.

Elder, K., Dale, B., 2019. Stem cell biology. In: In-Vitro Fertilization. Cambridge University Press, pp. 135–155, https://doi.org/10.1017/9781108611633.008.

Fagan, M., 2019. Stem cell biology. In: The Matrix of Stem Cell Research. Routledge, pp. 111–130, https://doi.org/10.4324/9781315104386-8.

George, J., Manjusha, W.A., Jegan, S.R., Mahija, S.P., Josphin, J.S., 2018. A review of stem cells in regenerative medicine. Int. J. Sci. Res. Sci. Technol. 3, 806–815.

Grochowski, C., 2018. Neural stem cell therapy-brief review. Clin. Neurol. Neurosurg. 173, 8–14. https://doi.org/10.1016/j.clineuro.2018.07.013.

Guo, Y., Pu, W.T., 2020. Cardiomyocyte maturation: new phase in development. Circ. Res. 126 (8), 1086–1106. https://doi.org/10.1161/CIRCRESAHA.119.315862.

Gurusamy, N., Alsayari, A., Rajasingh, S., Rajasingh, J., 2018. Adult stem cells for regenerative therapy. Prog. Mol. Biol. Transl. Sci. 160, 1–22. https://doi.org/10.1016/bs.pmbts.2018.07.009.

Jeeva Bharathi, S., Prasath, R.S.H., Sriram, K., Shankar, A., 2019. A review on regenerative braking system. Int. J. Psychosoc. Rehabil. 23, 183–193. https://doi.org/10.37200/IJPR/V23I3/PR190119.

Kaufman, D., Thomson, J., 2019. Embryonic Stem Cells. pp. 5–10, https://doi.org/10.1201/9780367813895-2.

Klimanskaya, I., Chung, Y., Becker, S., Lu, S.-J., Lanza, R., 2006. Human embryonic stem cell lines derived from single blastomeres. Nature 444, 481–485. https://doi.org/10.1038/nature05142.

Kolagar, T.A., Farzaneh, M., Nikkar, N., Khoshnam, S.E., 2020. Human pluripotent stem cells in neurodegenerative diseases: potentials, advances and limitations. Curr. Stem Cell Res. Ther. 15, 102–110. https://doi.org/10.2174/1574888X14666190823142911.

Kubota, H., 2018. Spermatogonial stem cells. Biol. Reprod. 99 (1), 52–74. https://doi.org/10.1093/biolre/ioy077.

Laplane, L., Solary, E., 2019. Towards a classification of stem cells. elife 8, e46563. https://doi.org/10.7554/eLife.46563.

Le, R., Huang, Y., Zhao, A., Gao, S., 2020. Lessons from expanded potential of embryonic stem cells: moving toward totipotency. J. Genet. Genomics 47, 123–130. https://doi.org/10.1016/j.jgg.2020.02.003.

Lenkiewicz, A.M., 2019. Epidermal stem cells. Adv. Exp. Med. Biol. 1201, 239–259. https://doi.org/10.1007/978-3-030-31206-0_12.

Liu, G., David, B.T., Trawczynski, M., Fessler, R.G., 2020a. Advances in pluripotent stem cells: history, mechanisms, technologies, and applications. Stem Cell Rev. Rep. 16, 3–32. https://doi.org/10.1007/s12015-019-09935-x.

Liu, S., Liu, F., Zhou, Y., Jin, B., Sun, Q., Guo, S., 2020b. Immunosuppressive property of MSCs mediated by cell surface receptors. Front. Immunol. 11, 1076. https://doi.org/10.3389/fimmu.2020.01076.

Martinez-Gamero, C., Malla, S., Aguilo, F., 2021. LSD1: expanding functions in stem cells and differentiation. Cells 10 (11), 3252. https://doi.org/10.3390/cells10113252.

McQualter, J.L., 2019. Endogenous lung stem cells for lung regeneration. Expert Opin. Biol. Ther. 19 (6), 539–546. https://doi.org/10.1080/14712598.2019.1596256.

Mobley, A., 2019. Induced Pluripotent Stem Cells. pp. 67–94, https://doi.org/10.1016/B978-0-12-811014-0.00004-4.

Mora, C., Serzanti, M., Consiglio, A., Memo, M., Dell'Era, P., 2017. Clinical potentials of human pluripotent stem cells. Cell Biol. Toxicol. 33, 351–360. https://doi.org/10.1007/s10565-017-9384-y.

Muhammad, M., 2020. Human Embryonic Stem Cells. Springer.

Nurković, J.S., et al., 2020. Corneal stem cells as a source of regenerative cell-based therapy. Stem Cells Int., 8813447. https://doi.org/10.1155/2020/8813447.

Pearson, R.G., Bhandari, R., Quirk, R.A., Shakesheff, K.M., 2017. Recent advances in tissue engineering. J. Long-Term Eff. Med. Implants 27, 199–231. https://doi.org/10.1615/JLongTermEffMedImplants.v27.i2-4.70.

Peters, E.B., 2018. Endothelial progenitor cells for the vascularization of engineered tissues. Tissue Eng. Part B Rev. 24 (1), 1–24. https://doi.org/10.1089/ten.TEB.2017.0127.

Poulos, J., 2018. The limited application of stem cells in medicine: a review. Stem Cell Res. Ther. 9 (1). https://doi.org/10.1186/s13287-017-0735-7.

Rajan, P., 2014. Human pluripotent stem cells. Ref. Modul. Biomed. Sci. https://doi.org/10.1016/B978-0-12-801238-3.00057-X.

Ramos, T., Moroni, L., 2020. Tissue engineering and regenerative medicine 2019: the role of biofabrication-a year in review. Tissue Eng. Part C Methods 26, 91–106. https://doi.org/10.1089/ten.TEC.2019.0344.

Riveiro, A.R., Brickman, J.M., 2020. From pluripotency to totipotency: an experimentalist's guide to cellular potency. Development 147. https://doi.org/10.1242/dev.189845.

Saeedi, P., Halabian, R., Imani Fooladi, A.A., 2019. A revealing review of mesenchymal stem cells therapy, clinical perspectives and modification strategies. Stem Cell Investig. 6, 34. https://doi.org/10.21037/sci.2019.08.11.

Samadi, P., Saki, S., Manoochehri, H., Sheykhhasan, M., 2021. Therapeutic applications of mesenchymal stem cells: a comprehensive review. Curr. Stem Cell Res. Ther. 16, 323–353. https://doi.org/10.2174/1574888X15666200914142709.

Sameri, S., Samadi, P., Dehghan, R., Salem, E., Fayazi, N., Amini, R., 2020. Stem cell aging in lifespan and disease: a state-of-the-art review. Curr. Stem Cell Res. Ther. 15, 362–378. https://doi.org/10.2174/1574888X15666200213105155.

Seiler, A.E.M., Spielmann, H., 2011. The validated embryonic stem cell test to predict embryotoxicity in vitro. Nat. Protoc. 6, 961–978. https://doi.org/10.1038/nprot.2011.348.

Suman, S., Domingues, A., Ratajczak, J., Ratajczak, M.Z., 2019. Potential clinical applications of stem cells in regenerative medicine. Adv. Exp. Med. Biol. 1201, 1–22. https://doi.org/10.1007/978-3-030-31206-0_1.

Sundaravadivelu, P.K., Raina, K., Thool, M., Ray, A., Joshi, J.M., Kaveeshwar, V., et al., 2021. Tissue-restricted stem cells as starting cell source for efficient generation of pluripotent stem cells: an overview. Adv. Exp. Med. Biol. https://doi.org/10.1007/5584_2021_660.

Swamy, V.E., et al., 2021. Growth kinetics and phenotypic markers expression in human dermal stem cells. Res. J. Biotechnol. 16 (12), 24–29. https://doi.org/10.25303/1612rjbt2429.

Tsutsui, T.W., 2020. Dental pulp stem cells: advances to applications. Stem Cells Cloning 13, 33–42. https://doi.org/10.2147/SCCAA.S166759.

Um, S., 2020. Prospects for the therapeutic development of umbilical cord blood-derived mesenchymal stem cells. World J. Stem Cells 12 (12), 1511–1528. https://doi.org/10.4252/wjsc.v12.i12.1511.

Xu, M., He, J., Zhang, C., Xu, J., Wang, Y., 2019. Strategies for derivation of endothelial lineages from human stem cells. Stem Cell Res Ther 10, 200. https://doi.org/10.1186/s13287-019-1274-1.

Yamanaka, S., 2020. Pluripotent stem cell-based cell therapy-promise and challenges. Cell Stem Cell 27, 523–531. https://doi.org/10.1016/j.stem.2020.09.014.

Zaehres, H., 2020. Induced pluripotent stem cells. In: Essential Current Concepts in Stem Cell Biology. Springer, pp. 123–136, https://doi.org/10.1007/978-3-030-33923-4_7.

Zakrzewski, W., Dobrzyński, M., Szymonowicz, M., Rybak, Z., 2019. Stem cells: past, present, and future. Stem Cell Res Ther 10, 68. https://doi.org/10.1186/s13287-019-1165-5.

Zhang, Q., Yu, J., Chen, Q., Yan, H., Du, H., Luo, W., 2021. Regulation of pathophysiological and tissue regenerative functions of MSCs mediated via the WNT signaling pathway. Mol. Med. Rep. 24. https://doi.org/10.3892/mmr.2021.12287.

Zhao, Q., et al., 2020. Intestinal stem cells and intestinal organoids. J. Genet. Genomics 47 (6), 289–299. https://doi.org/10.1016/j.jgg.2020.06.005.

CHAPTER

2 Animal stem cells: Extraction, expansion, and cryopreservation

Sakinah-Syed[a] and Suresh Kumar Subbiah[b]

[a]*Department of Medical Microbiology, Faculty of Medicine and Health Sciences, Universiti Putra Malaysia, Serdang, Selangor, Malaysia,* [b]*Centre for Materials Engineering and Regenerative Medicine, Bharath Institute of Higher Education and Research, Chennai, Tamil Nadu, India*

1. Introduction

Stem cells have become a revolution in the medicine and biology world. Many animal studies have become a fundamental baseline for developing successful human trials. Although there are developing human trials using human stem cell lines to treat diseases, more animal stem cell research is required due to equivocal results from the human trials. Stem cells are universally classified into two main categories: embryonic stem cells (ESCs) and adult stem cells (ASCs) (Fortier and Travis, 2011).

1.1 Embryonic stem cells (ESCs)

Stem cells can be found at all stages of life; however, scientists have been focusing on three main stages of stem cell development, which are ESCs, fetal stem cells, and ASCs. Due to the multidifferentiation ability of ESCs, they are widely used in the field of regenerative medicine. ESCs can be obtained from the inner layer of approximately 5-day-old embryos, which consist of two-layered ball of cells. Amazingly, ESCs can also be created through test-tube fertilization and implanted into different species or animals without facing autoimmune rejection. Although, its clinical application is controversial due to ethical issues and likelihood of carcinogenesis (Guasch and Fuchs, 2005).

1.2 Adult stem cells (ASCs)

Unlike ESCs, ASCs are not pluripotent and are harder to isolate due to their minute number. ASCs are found in various human tissues, mainly in the bone marrow and in the brain, blood, liver, skin, and pancreas, which can only generate specific types of

Stem Cell Laboratory Techniques. https://doi.org/10.1016/B978-0-12-823729-8.00002-6

cells. For example, HSCs derived from bone marrow can only differentiate into blood cells (lymphocytes, macrophages, red blood cells, and platelets), but it is difficult to differentiate into liver, muscle, or neuron cells. Despite these limitations, ASCs have attracted attention for vast clinical trials such as auto-grafting cells due to their ability to differentiate into various cell types. Meanwhile, fetal stem cells are obtained from placenta near umbilical cord. The samples collected from this area are relatively small; however, it is possible to make the stem cells multiply (Guasch and Fuchs, 2005).

In this chapter, we will discuss the most commonly used techniques to isolate these stem cells from laboratory animals, which will be fundamental guidelines for researchers to conduct their animal-derived stem cell research. Mice were used to explain most of the techniques; however, these procedures are applicable to various rodent laboratory animals.

2. Materials, reagents, chemicals, apparatus, and kits

2.1 Embryonic stem cells (ESCs) isolation

1. Pregnant mouse at 3.5 days postcoitum (d.p.c.)
2. Mouse embryonic fibroblasts (MEF) feeder cells
3. Knockout serum replacement (KOSR) ES medium (filter the medium using 0.22 μm filter)
   a. Dulbecco's modified Eagle's medium (DMEM) high glucose no sodium pyruvate, no glutamine (Invitrogen, New York, United States)
   b. 15% KnockOut serum replacement (KOSR; Invitrogen, New York, United States)
   c. 1× penicillin-streptomycin (Invitrogen, New York, United States)
   d. 2 mM GlutaMAX (Invitrogen, New York, United States)
   e. 1 mM sodium pyruvate (Invitrogen, New York, United States)
   f. 0.1 mM MEM nonessential amino acid solution (NEAA; Invitrogen, New York, United States)
   g. 0.1 mM 2-mercaptoethanol (Invitrogen, New York, United States)
   h. 10^3 IU of mouse leukemia inhibitory factor (LIF; Millipore, Hassen, Germany)
   i. 1 μM PD0325901 (MEK inhibitor; Stemgent, Maryland, United States)
   j. 3 μM CHIR99021 (glycogen synthase kinase [GSK] 3 inhibitor; Stemgent, Maryland, United States)
4. Fetal bovine serum (FBS) ES medium (filter the medium using 0.22 μm filter)
   a. DMEM high glucose no sodium pyruvate, no glutamine
   b. 15% FBS ES cell grade, filtered in 0.45 μm filter (Lonza, California, United States)
   c. 1× penicillin-streptomycin
   d. 2 mM GlutaMAX

   e. 1 mM sodium pyruvate
   f. 0.1 mM MEM NEAA
   g. 0.1 mM 2-mercaptoethanol
   h. 10^3 IU of LIF
   i. 1 μM PD0325901
   j. 3 μM CHIR99021
5. Serum-free ES medium (filter the medium using 0.22 μm filter)
   a. 1 Part of DMEM/F-12 medium (Invitrogen, New York, United States) supplemented with N-2 (Invitrogen, New York, United States)
   b. 1 Part of neurobasal (Invitrogen, New York, United States) supplemented with B-27 (Invitrogen, New York, United States)
   c. Mix equal part of medium from 5a and 5b
   d. Similar number of reagents in 4 c-j is added to the serum-free ES medium to prepare a complete culture medium
6. MEF medium
   a. DMEM high glucose no sodium pyruvate, no glutamine
   b. 10% FBS
   c. 1× Penicillin-streptomycin
   d. 2 mM GlutaMAX
7. 3 μM CHIR99021 stock preparation
   a. Resuspend 2 mg of CHIR99021 in 140 μL of dimethyl sulfoxide (DMSO; Sigma-Aldrich, Hassen, Germany)
8. 1 μM PD0325901 stock preparation
   a. Resuspend 2 mg of PD0325901 in 414 μL of DMSO
9. 1 mg/mL Mitomycin C in PBS (Sigma-Aldrich, Hassen, Germany)
10. Defined trypsin inhibitor (Invitrogen, New York, United States)
11. 0.05% Trypsin-EDTA (Invitrogen, New York, United States)
12. 70% Alcohol (Sigma-Aldrich, cat. no. E7023)
13. 1× EmbryoMax M2 media (Millipore, Hassen, Germany)
14. Phosphate-buffered saline (PBS) without calcium and magnesium (Invitrogen, New York, United States)
15. Acid Tyrode's solution (Sigma-Aldrich, Hassen, Germany)
16. 0.1% Gelatin (STEMCELL Technologies, Biopolis, Singapore)

2.2 Mesenchymal stem cells (MSCs) isolation from umbilical cord

1. 8-week-old Kunming mice
2. PBS (Invitrogen, New York, United States)
3. DMEM/F-12 nutrient medium mixture (Hyclone Laboratories, Utah, United States)
4. FBS (Gibco, Massachusetts, United States)
5. Penicillin-streptomycin (Invitrogen, New York, United States)
6. Culture medium recipe

a. DMEM/F-12 nutrient medium mixture supplemented with 15% FBS and 100 U penicillin-streptomycin
7. 0.25% Trypsin-EDTA (Invitrogen, New York, United States)

2.3 Stem cell isolation from bone marrow

1. 0.25% trypsin-EDTA (Invitrogen, New York, United States)
2. MEMalpha powder with nucleosides (or DMEM) adjusted to pH7.2 and filtered for sterilization
   a. One pack of MEM-alpha powder for 1 L (Life Technologies, PA4 9RF, United Kingdom)
   b. 2.2 g $NaHCO_3$ (Sigma-Aldrich, Missouri, United States)
   c. 1 L of Milli-Q water
3. Complete MEM-alpha (or DMEM) medium
   d. MEM-alpha stock medium
   e. 15% Heat-inactivated FBS (Life Technologies, PA4 9RF, United Kingdom)
   f. 1% Penicillin-streptomycin-neomycin solution (PSN; Life Technologies, California, United States)
4. PBS adjusted to pH7.4 and sterilized
   g. 8.0 g NaCl
   h. 1.44 g Na_2HPO_4
   i. 0.2 g KCl
   j. 0.24 g KH_2PO_4
   k. 1 L Milli-Q water

2.4 Isolation of HSCs derived from bone marrow or umbilical cord stem cells

1. Suggestions: BD hematopoietic stem and progenitor cell isolation, EasySep hematopoietic progenitor isolation kit, CD117-positive selection kits, MACS cell separation technology or lineage cell depletion kit.
2. 0.22 μm sterile filtered FACS buffer
   a. 0.1 g bovine serum albumin (BSA; Sigma-Aldrich, Missouri, United States)
   b. 100 mL PBS
3. Trypan blue stain
4. Hoechst/propidium iodide/7-AAD (7-aminoactinomycin D; Invitrogen, New York, United States)
5. Fc blocker antibody (antimouse CD16/32; eBioscience, California, United States)
6. Isotype controls (mIgG1 tagged FITC/PE/APC; eBioscience, California, United States)
7. Antimouse lineage cocktail-FITC antibody (eBioscience, California, United States)

8. Antimouse Sca-1-APC antibody (eBioscience, California, United States)
9. Antimouse c-Kit-PE antibody (eBioscience, California, United States)
10. DMEM high glucose (Invitrogen, New York, United States)
11. FBS (Gibco, Massachusetts, United States)
12. Penicillin-streptomycin (Invitrogen, New York, United States)
13. Murine SCF growth factor (Biovision, California, United States)
14. Murine IL-3 growth factor (Biovision, California, United States)
15. Murine IL-6 growth factor (Biovision, California, United States)
16. 70% Alcohol
17. Round bottom polypropylene FACS tubes
18. High-speed cell sorter (BD FACS Aria or BD influx)

2.5 Stem cell isolation from peripheral blood

1. 6- to 8-week-old mice
2. Iscove's modified Dulbecco's medium (IMDM; Gibco, Lowland, Scotland)
3. 5% Fetal calf serum (FCS; Invitrogen, California, United States)
4. 25 IU Heparin (Leo Pharma, North Brabant, The Netherlands)
5. Lympholyte-M (Cedarlane Laboratories Ltd, Hornby, Canada)
6. Biotinylated lineage-specific mouse lineage panel antibodies (anti-CD45R, anti-CD11b, anti-TER119, anti-Gr-1, and anti-CD3e; BD Pharmingen, California, United States)
7. c-Kit-APC antibody (BD Pharmingen, California, United States)
8. Sca-1-FITC antibody (BD Pharmingen, California, United States)
9. Streptavidin-phycoerythrin (PE; Pharmingen, California, United States)
10. Sterile FACS buffer

2.6 Stem cell isolation from adipose tissue

1. 6- to 8-week-old mice
2. 70% Alcohol
3. PBS
4. Digestion medium
    a. 2 mg/mL collagenase type I in PBS for 1 g fat. Collagenase type VIII also can be used as an alternative for more stromal vascular fraction (SVF) cells. (Worthington Biochemical Corporation, New Jersey, United States)
5. $CaCl_2$
6. Complete medium
    a. DMEM/F-12 medium (Invitrogen, California, United States)
    b. 10% FBS (heat inactivated; Thermo Fisher Scientific, Massachusetts, United States)
    c. 1% Penicillin-streptomycin (Invitrogen, California, United States)
    d. Suggestion: Add basic FGF (5 ng/mL; Sigma-Aldrich, Missouri, United States) to enhance the growth of mouse ADSCs during expansion.

7. 50–70 μm cell strainer
8. 10 cm collagen coated plate

2.7 Neural stem cell isolation

1. 5- to 8-month-old adult mouse
2. Equipment and reagents for euthanasia
3. 70% Alcohol.
4. Papain dissociation kit (Worthington Biochemical Corporation, New Jersey, United States)
5. DMEM/F-12 medium (Thermo Fisher Scientific, New York, United States)
6. Neural stem cell (NSC) medium
    a. 500 L DMEM/F-12 (Thermo Fisher Scientific, New York, United States)
    b. 10 mL from 50× B-27 supplement (Thermo Fisher Scientific, New York, United States)
    c. 5 mL from 100× antibiotic-antimycotic (Thermo Fisher Scientific, New York, United States)
    d. 6.75 mL from 100× Glutamax (Thermo Fisher Scientific, New York, United States)
7. 20 μg/mL EGF (Peprotech, New Jersey, United States)
    a. Reconstitute 20 μg EGF in 1 mL sterile distilled water
8. 10 μg/mL FGF (Peprotech, New Jersey, United States)
    a. Reconstitute 10 μg FGF in 1 mL of 5 mM Tris (pH 7.6)
9. Glass pasteur pipettes and bulbs (Sterilized)
10. Incubator (37°C, 5% CO_2 humidity) with mechanical shaker
11. High-speed table-top centrifuge
12. Nonadherent cell culture flask
13. Trypan blue
14. Accutase
15. PBS
16. Poly-ornithine or fibronectin coated plates

2.8 Stem cell isolation from epidermis

1. 2- to 3-month-old male Balb/c
2. 100% carbon dioxide (CO_2) with euthanizing chamber
3. 70% Alcohol
4. 1 mg/mL Liberase (Roche, Baden-Württemberg, Germany) in Hanks' Balanced Salt Solution (HBSS)
5. Dispase 100-caseinolytic U/mL (Life Technologies, California, United States)
6. DMEM (Gibco, Maryland, United States)
7. 10% FBS (Gibco, Maryland, United States)
8. Trypan blue
9. Collagen IV (R&D Systems, Minnesota, United States) to coat the culture plates

10. Skin-derived precursors (SKP) culture medium
    a. DMEM/F-12 at a ratio of 3:1 (Gibco, Maryland, United States)
    b. 20 ng/mL EGF (R&D Systems, Minnesota, United States)
    c. 40 ng/mL bFGF (R&D Systems, Minnesota, United States)
    d. 1% B-27 supplement (Invitrogen, Massachusetts, United States)
11. Mesenchymal culture medium
    a. DMEM (Gibco, Maryland, United States)
    b. 10% FBS (Gibco, Maryland, United States)
    c. 1% Ampicillin-streptomycin (Invitrogen, California, United States)
12. EpiSC culture medium
    a. Keratinocyte serum free (KSF) medium (Gibco, Maryland, United States)
    b. EPB supplements
    c. 20 ng/mL EGF (R&D Systems, Minnesota, United States)
    d. 1% Ampicillin-streptomycin (Invitrogen, California, United States)
13. 0.1% Trypsin

2.9 Cryopreservation

1. 0.25% trypsin-EDTA
2. PBS
3. Cell culture complete medium (specific media for particular cell type) supplemented with 5%–10% FBS (concentration based on specific cell type used)
4. Tryphan blue
5. Freezing media (10% DMSO and 90% FBS or 70% basal medium, 20% FBS and 10% DMSO)
6. Freezing container containing 100% isopropyl alcohol
7. Liquid nitrogen tank

3. Procedure

3.1 Embryonic stem cells (ESCs) isolation

ESCs are pluripotent cells that are derived from inner cell masses (ICMs) that have the potential for self-renewal and give rise to all types of body cells. The existence of stem cells and ESCs was first and foremost revealed through animal research allowing researcher to better understand ESCs' characters and peculiarity including improvising their methods. This is done with species that are less related or closely related to humans. Eventually, human ESCs were established in 1998; however, debate and controversies are still going on about the status and legality of using human embryo, which may hamper their development in human studies. The ESCs' research through nonhuman derivatives has helped and improved a lot in stem cell clinical therapy; however, some aspects still need to be addressed. Hence, studies using ESCs are still a necessity. Here, we will discuss on how to establish mouse

ESC lines. This method can also be applied using other rodent laboratory animal with some modification required on the chemicals and reagent that will be used (Jozefczuk et al., 2012; Lee, 2013).

1. Inactivation of MEF
   a. Culture MEFs until reaching confluency in MEF medium.
   b. Replenish current medium with fresh medium supplemented with 10 μg/mL mitomycin C.
   c. Incubate for 2–3 h followed by washing with PBS for several time.
   d. Collect the cells by trypsinization (0.05% trypsin-EDTA).
   e. Cryopreserve the cells or
   f. Plate 1×10^5 cells/cm^2 on a gelatinized plate in 100 mL of MEF medium. This procedure is carried out 1 day before the collection of blastocysts.
   g. The next day, wash the plates several time with PBS and replace with either serum-free ES medium (for blastocysts from nonpermissive strain) or KOSR ES medium (for blastocysts from permissive strain) on the day of blastocysts extraction. Place the plates in the incubator until the blastocysts is ready for plating.
2. Euthanize a pregnant mouse is by cervical dislocation at 3.5 d.p.c. (Fig. 1).
3. Dissect the uterine horns. Trim excessive fat and then rinse the uteri horn in a falcon tube containing 70% alcohol in short.
4. Transfer the uterine horn into magnesium/calcium-free PBS and bring the tube into tissue culture biosafety cabinet to proceed with aseptic techniques.
5. Place the uterine horns on a petri dish and secure it to a syringe (containing M2 medium) using forceps.
6. Flush out the blastocysts with 2 mL of M2 medium using 5 mL syringe attached to 27 G needle (other suggestions: 1, use transfer straw; 2, cut open the uterine horn).
7. Wash the flushed blastocysts with few drops of M2 medium by using any suitable micropipetting aid or a mouth pipette.
8. By using drilling micropipette, dissolve the blastocysts by incubating the blastocysts in 0.5 mL microdroplets of acid Tyrode's solution up to 30 s to circumferentially dissolve the outer layer of zona pellucida. This procedure needs to be done rapidly to avoid exposure of the blastocysts to the acidic solution.
9. Observe the blastocysts under inverted microscope. If the zona is not dissolved, blastocyst can be further incubated in the solution for another 15–30 s. Wash the blastocysts carefully several time with PBS.
10. Plate a fully expanded zona pellucida-free blastocysts using either serum-free ES medium (for blastocysts from nonpermissive strain) or KOSR ES medium (for blastocysts from permissive strain) according to the requirements of the experiment. It is recommended to seed one blastocyst per well of a 4-well plate. This plate is seeded with mitomycin C-inactivated MEF feeder cells 1 day before the fully expanded blastocysts are seeded.

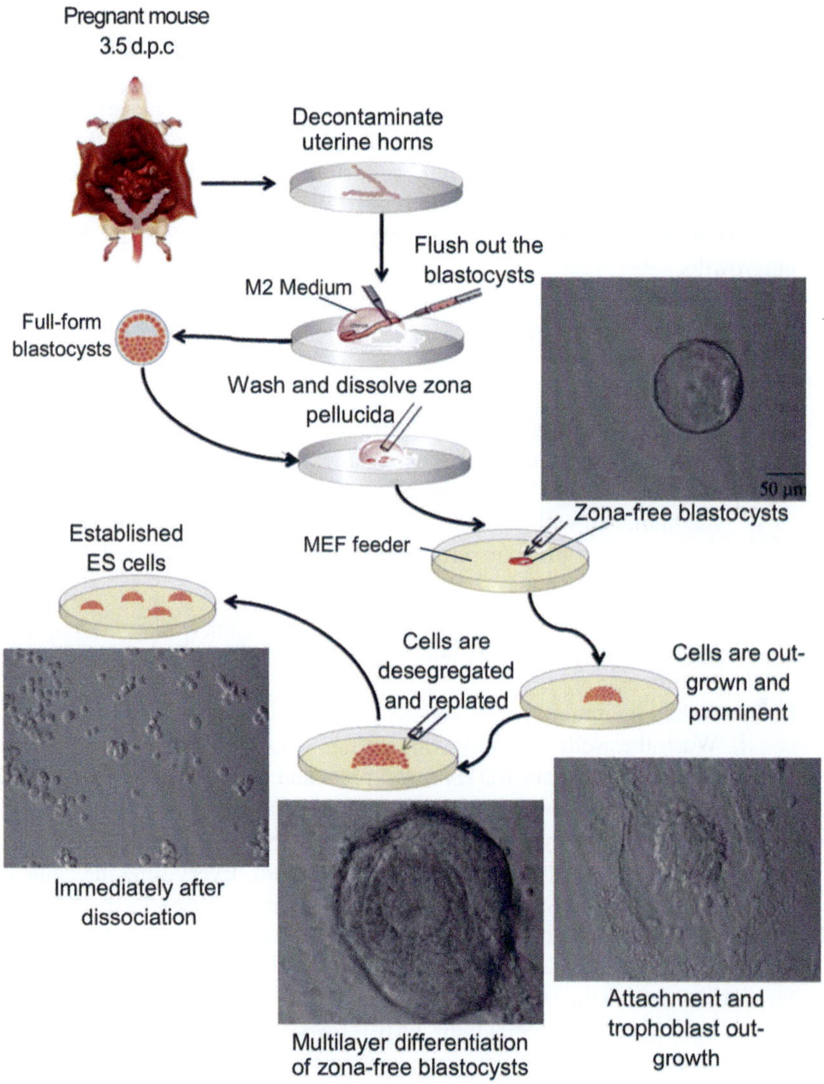

FIG. 1

Mouse embryonic stem cell isolation.

Microscopic images adopted from Tanaka, N., Takeuchi, T., Neri, Q.V., et al., 2006. Laser-assisted blastocyst dissection and subsequent cultivation of embryonic stem cells in a serum/cell free culture system: applications and preliminary results in a murine model. J. Transl. Med. 4, 20. https://doi.org/10.1186/1479-5876-4-20.

11. Incubate the plated blastocysts are for 48 h without any disturbance at 37°C, 5% CO_2 incubator. At this stage, the blastocysts will hatch and attach to the feeder cells.
12. Change half of the media on the third day of incubation. Replace half of the media every day until the cells are prominent (three to four times the size of blastocysts). At this timepoint, the cells are ready for disaggregation. The growth of the cells should be monitored every day from the third day of culture to prevent from overgrowth such as very large outgrowths or dark-centered outgrowths.
13. The outgrowths are dispersed carefully by using either:
    a. Mechanical treatment:
        i. The outgrowth is transferred to a new MEF feeder cell-covered 4-well plates.
        ii. Break up the outgrowth using sterile tweezers into smaller pieces.
        iii. Pipette the fragments for several time.
        iv. Replace with new growth medium. If the cells were grown in KOSR ES medium, cell will be further grown in FBS ES cell medium. However, if the cells were grown in serum-free ES, the same medium will be used for further experiment.
        v. Enzymatically passage the colony once the colony emerged approximately 2–3 days later at 1:1 ratio as described in enzymatic treatment.
    b. Enzymatic treatment:
        i. Wash the wells carefully with PBS.
        ii. Add 0.05% trypsin-EDTA to the plate and incubated at 37°C for 5 min.
        iii. Pipette the fragments vigorously for several time.
        iv. Inactivate the trypsin by adding defined trypsin inhibitor if the outgrowth is cultured in serum-free medium. Every time the cells are exposed to trypsin, the passage number are increased by 1.
14. Replace the medium with appropriate growth medium (FBS ES cell medium or serum-free ES medium) and transfer to new MEF feeder cell-covered 4-plate.
15. Passage the cells at the ratio of 1:4 (seeding density is ~$3–5 \times 10^5$ cells per well or ~$1–4 \times 10^5$ cells per cm^2) onto a MEF feeder cell-covered 6-well plate or 35-mm dish.
16. Observe the cells on daily basis and replace the medium accordingly or proceed with desired experiment or cryopreservation.

3.2 Stem cell isolation from umbilical cord

In current decade, umbilical cord-derived stem cells (UC-SCs) are gaining attention for its therapeutic potential. Some of the reason includes lower rejection incidence during transplantation, does not require ideal antigen matching as in BM transplantation, and contains more primary HSCs per volume compared with BM. Apart from that, hospitals have had an established method for collecting, storing, and freezing the UC since the 1940s, which does not require any new technology. The UC is a home for pluripotent cells, HSCs, and mesenchymal-like stem cells. Interestingly, UC-SCs have more fascinating properties compared with stem cell derived from other tissue sources such as their genetic, cellular, allogeneic, and xenogeneic treatment variations. They also appear to have higher expansion capacity, greater proliferation potential, and shorter doubling time compared with SCs from distinct tissue sources. Successful isolation of UC-SCs has been done from humans, horses, cows, pigs, and chickens; however, very few methods have been established for UC-derived stem cells in rodents. It is important to study on rodent primary cells to elucidate the molecular mechanisms of diseases in basic and preclinical studies. Hence, we share an easy technique to obtain stem cells from mouse UCs to support in vivo experiments (Li et al., 2013; Zhang et al., 2018).

1. A Kunming mice is used to describe the experimental procedures.
2. At the gestational age of 15–19 days of Kunming mice, collect the fresh mouse umbilical cords (UCs) using aseptic techniques.
3. Rinse the collected UCs with PBS containing 1× penicillin-streptomycin.
4. Place the UCs on a petri dish and mechanically cut the washed tissue into small (0.5-1.0 mm^3) sections
5. Suspend the tissue in culture medium
6. Seed the suspended tissue with the culture medium into 3.5 mm cell culture dishes.
7. Incubate the dish in humidified 5% CO_2 incubator at 37°C.
8. Replace the medium every 3 days once to remove all the cell debris and nonadherent tissues.
9. Heterogeneous primary cell colonies will be well developed around 9–12 days. The cell now can be trypsinized for first passage using 0.25% trypsin-EDTA.
10. It is suggested to perform the first passage at the ratio of 4:1–2:1, and the consecutive passages (passages 3–5) at the ratio of 1:1–1:2. At this stage, medium is recommended to be replaced every 2 days.
11. The homogenously fibroblast-like cell populations will be visible after passage 5.
12. UM-MSCs can be obtained at passage 7–12 for further experiments.

3.3 Stem cell extraction from bone marrow

Like umbilical cord, bone marrow (BM) also contains a mixture of stem cells (mononucleated cells—MNCs), which are endothelial cell progenitors, HSCs, and MSCs. These stem cells can differentiate into mature cells such as blood cells, fat, bone, cartilage, epithelial cells of lung, liver, kidney, gastrointestinal tract skin, and myocytes of skeletal and heart muscles. Thus, BM-derived stem cells are an important source for cell-based therapy. Many clinical trials have been done using stem cells from BM sources because they are easy to collect and require minimal manipulation, be it immunological or density gradient separation. Below, easy-to-perform procedures will help researcher to isolate and culture mouse BM-MNCs (Astori et al., 2010; Herzog et al., 2003; Huang et al., 2015; Sakinah et al., 2021).

1. To extract the bone marrow cells, euthanize the donor mice/rat by cervical dislocation, CO_2 exposure or by overdose of anesthetics chemicals.
2. Clean the mice with 70% alcohol, and carry out the further procedure in laminar flow (Fig. 2A).
3. Remove the skin of all four limbs using a pair of scissors to isolate the bone marrow from both hind limb femurs and tibias.
4. Dissect all four limbs at the carpel and ankle joints by making incisions around the joint between trunk and hind limbs, trunk, and forelimbs (Fig. 2B–D).
5. Carefully remove the muscles, tendons, and ligaments from tibias, femurs, and humeral bones using surgical scalpel and microdissecting scissors, while residual muscles can be completely removed using gauze. Remove the soft tissues completely from the bones to avoid cross contamination (Fig. 2E–G).
6. Soak the bones in 70% alcohol for less than 1 min, then wash in PBS thrice, and finally soak into MEM alpha or DMEM supplemented with 2% FBS before separating the femur and tibia.
7. To expose the bone marrow cavity, make incision in both ends of epiphysis of bones using scissors without damaging the bone structure (Fig. 2H).
8. Aspirate DMEM media supplemented with 2% FBS into a 26 G syringe and puncture into the bone marrow cavity and flush the media out through the bones until the reddish line in the bone become invisible. This step can be done in petri dish or in a 50 mL tubes for succeeding processes (Fig. 2I–K).
9. To ensure high cell viability, it is suggested to process all the samples within 30 min.
10. To obtain MSCs, first resuspend the cavity with medium to dissociates the cells and proceed with centrifugation at $300 \times g$ for 10 min. After that, seed the cells in petri dish or T-25 flask in DMEM high glucose medium, supplemented with 20% FBS, 1% penicillin-streptomycin, and 1% glutamine. Replace the media every 2–3 days.
11. It is suggested to passage at a ratio of 1:3. The BM-MSCs can be used for experiment after passage 3 as the culture contains fewer blood cells, fat, and macrophages.

3.4 Isolation of HSCs derived from bone marrow or umbilical cord stem cells

Hematopoietic stem cells (HSCs) are multipotent cells that differentiate into blood cell lineage. The main characteristic of HSCs is its ability to restore the complete blood cell lineages after ablation of bone marrow. Thus, they are very crucial and ideal target for diverse clinical application such as gene therapy and SC transplantation. In most clinical setting, HSCs isolated from cord blood or peripheral blood is easier because it is easier to obtained when compared to bone marrow derived HSCs. However, due to very low yield, it is difficult to isolate from small laboratory animals. It is estimated that 1 HSCs can be obtained from 10,000 bone marrow cells, but only 1 HCSs is found in every 100,000 blood cells. Therefore, HSC isolation mainly

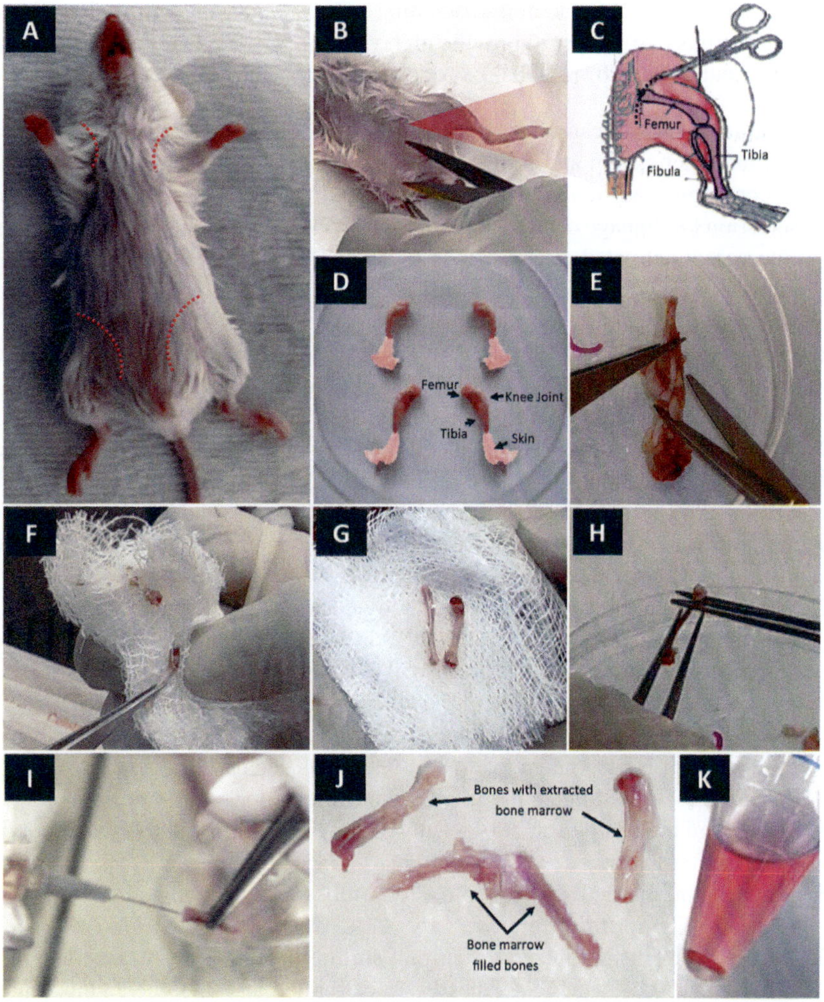

FIG. 2

Illustration of the bone marrow isolation from mouse limb bones. (A) Euthanize the mouse and decontaminate with 70% alcohol. (B, C) Dissect all leg limbs from the hip bones. (D) Remove skin and separates all the bones, incision can be made at the knee joint. (E) Separated the muscles, tendons, and ligaments from tibias, femurs, and humeral bones and (F, G) completely remove the residual muscles using gauze. (H) Make incision in both ends of epiphysis of bones and (I) flush out the bone marrow cavity using syringe. (J) A clear bone or white color bones will be seen as an indication that the bone marrow cavity is successfully extracted out. (K) A reddish color sediment also can be observed in the collection tube.

from bone marrow (BM) and umbilical cord (UC) derived stem cells is chosen to be discussed in this part aiming to provide high yield of HSCs (Kumar and Madabushi, 2018; Sakinah et al., 2021; Vadakke-Madathil et al., 2019).

1. Isolation of HSCs can be performed using Isolation kits such as BD Hematopoietic stem and progenitor cell isolation, MACS Cell Separation Technology, EasySep hematopoietic progenitor isolation or CD117-positive selection kit, lineage cell depletion kit, or by using traditional isolation method, fluorescent activated cell sorting (FACS) of HSCs. As the protocol for manufacturer's developed kits are available in their website, FACS technique to isolate HSCs specific for murine animal will be discussed here.
2. Suspend the cells (BM-SCs or UC-SCs) in sterile FACS buffer and perform cell counting using typan blue. Prepare cells by staining with fluorochrome-tagged antibodies prior to cell sorting as described below.
3. Aliquot 5×10^5 to 2×10^6 viable cells of BM-SCs or UC-SCs into each tube.
4. Add 1–2 μg Fc receptor-blocking antibody per 1×10^6 viable cells into each tube and incubate at 4°C for 10–15 min.
5. Without washing the cell suspension, proceed by adding lineage cell depletion cocktail and corresponding isotype controls to the samples.
6. Add optimized concentration of antibody cocktail (FITC-tagged lineage cocktail antibody, PE-tagged c-Kit and APC-tagged Sca-1 antibodies) to the cell suspension and incubate for 40 min at 4°C in dark. Mix the antibodies with cells by gentle intermittent flicking of the tubes. Fluorescent tags may be changed by the investigator as required by the experiment. After the incubation with antibodies, it is recommended to add vital dyes such as Hoechst/7-AAD/ propidium iodide, to gate the live cells during sorting setup.
7. Centrifuge the samples at $500 \times g$ at 4°C for 3 min followed by washing twice with FACS buffer.
8. Dispense the filtered samples into FACS tubes using 40 μm cell strainer and keep at 4°C prior to sorting.
9. The cell sorter is optimized for current sorting process. For instance, if user is using BD FACS Aria, CS&T beads are calibrated and then the gating is built using unstained cells, fluorescence minus one (FMO) control, isotype controls and positive control using BD FACSDiva software.
10. Apply gating for live cells using a free hand gating features followed by gating the Lin^- population. After that population. After that gate the populations that are double-positive Sca-1^+ (APC^+) and c-Kit^+ (PE^+) for sorting (sorting gate).
11. Sort all populations within the gate created (Lin^- c-Kit^+ $Sca1^+$) into the collecting FACS tube and followed by centrifugation.
12. Count the LSK progenitors and seed in T-25 flask at a density of 2×10^6 cells and Incubate at 37°C, 5% CO_2. The cells are maintained in high-glucose DMEM containing 15% FBS, 1% penicillin-streptomycin, 10 ng/mL human IL-6, 6 ng/mL murine IL-3, and 100 ng/mL murine SCF. The media is changed every 3 days by centrifugation at $200 \times g$ for 5 min prior to further experiments accordingly.

3.5 Stem cell isolation from peripheral blood

The use of peripheral blood stem cells is increasing in clinical transplant settings since past 2 decades. This is because the availability of fast extraction technique and regeneration of blood cells by peripheral blood stem cell compared with bone marrowderived stem cells. Peripheral blood stem cells speed up the hematological recovery, although their reduced potential for long-term repopulation of blood stem cells due to lower stem cell frequency in peripheral blood leads to the disadvantages of using this technique (Yeoh et al., 2007).

1. Harvest peripheral-derived stem cells via cardiac puncture under approval of IACUC animal handling approval.
2. Collect approximately 1 mL of blood, dilute the blood with 4 mL of IMDM containing 5% FCS and 25 IU heparin.
3. Now the total collected blood cell suspension is 5 mL.
4. Centrifuge the cell suspension over an equal volume of Lympholyte-M for 30 min in room temperature (25°C) at 400×*g*.
5. After that, isolate the opaque interface layer (mononuclear cells).
6. Resuspend the mononuclear cells in IMDM supplemented with 5% FCS and centrifuge at 2000 rpm for 5 min at 4°C.
7. Resuspend the cells in sterile FACS buffer, count the cells and prepare the cells for cell sorting with 5×10^5 to 1×10^6 viable cells per FACS tube.
8. Stain the cells with biotinylated mouse lineage specific cocktail antibody (anti-Gr-1, anti-TER119, anti-CD11b, anti-CD3e, and anti-CD45R), APC-anti-c-Kit and FITC-anti-Sca-1. Incubate for 40 min at 4°C in dark. Mix the antibodies with cells by gentle intermittent flicking of the tubes.
9. Visualize the biotinylated cocktail antibodies with streptavidin-phycoerythrin (PE).
10. After antibody staining, centrifuge the samples at 500×*g* at 4°C for 3 min followed by washing twice with FACS buffer.
11. Sort LSK and Lin^- non-Sca-1^+c-Kit^+ cells
12. Follow the procedure described in Section 3.4 (No. 8–11) and further use the cells as per experimental requirements.

3.6 Stem cell isolation from adipose tissue

Therapeutic studies on adipose-derived stem cells (ADSCs) are growing to be an important role in regenerative medicine. Studies have reported that stem cells that are derived from adipose tissue have a higher extraction rate compared with bone marrow-derived stem cells (BMSCs). Generally, adipocyte tissues have differentiation potential and plasticity, and are extracted from multipotent stem cells (ADSCs). Therapies based on ADSCs are important for numerous diseases and injuries as it helps in the functionality and as vascular building blocks for the damaged tissues. ADSCs hold a promising translational tissue repair abilities and are potential in clinical tissue regeneration therapies. Below describes detailed procedures of stem cell isolation from adipose tissues (Ferraro et al., 2016; Huang et al., 2013).

1. Euthanize 6- to 8-week-old mice and shave the fur clean.
2. Sterilize the with 70% alcohol by spraying it onto the mice fur before opening.
3. By holding the skin and fat, separate the mice dorsal fat layer with the skin layer using a scalpel from the muscles (Fig. 3).
4. Immediately immerse the fat layer attached to skin into a clean PBS, separate the subcutaneous inguinal WAT and interscapular BAT. (For interscapular BAT: Place the mouse with its back faced up, cut open the skin along the back until the neck. The interscapular (brown adipose tissue) is available under the skin between the shoulders. Meanwhile, BAT is visible as a butterfly shaped with two lobes. The BAT is covered with a thin layer of white fat and it should be removed carefully. For subcutaneous WAT, remove skin. Inguinal WAT is found under the skin on both sides, starting almost on the back until inside of the thighs and down toward the testicle.)
5. Remove all the contaminant from the collected adipose tissue such as skeletal muscle, excessive hair, contaminating blood, and connective tissues by washing them several times with PBS.
6. Mince the fat layer into small pieces and transfer the minced fat fragments into a 4 mL of digestion medium (or more following the weight of the mouse fat). Add $CaCl_2$ before digestion process to enhance the reaction (optional).
7. Pipette the minced fat up and down using a 5 mL pipette and then transfer the minced tissues in 50 mL tubes with the rest of the digestion medium, continue mixing for several times using pipette.
8. Incubate the fat tissue in digestion medium for 40 to 50 min in shaker, at 150 rpm with constant agitation in 37°C. Check the digestion every 10–15 min to avoid overdigestion.
9. Add 5 mL or equal amount of complete medium to the digestion medium to stop digestion and pipette up and down to mix well. (Make sure the cells are completely homogenous and do not have any clumping.)
10. Centrifuge it for 10 min at 700 × *g*.
11. Discard the oily mature adipocyte layer and the liquid layer, leaving the brownish pellet (consist of stromal vascular fractions (SVFs))
12. Dissolves and mix well the pellet in 10 mL complete medium.
13. Filter the cell suspension into a new 50 mL tube using a 50–70 μm diameter of cell strainer.
14. Transfer the cell suspension into a new 15 mL tube and proceed with centrifugation for 10 min at 700 × *g*.
15. Aspirate medium and resuspend the pellet in 10 mL complete medium. Pipette and mix well.
16. Prepare plates required to plate the cells. It is suggested that two plates of 10 cm is used to plate inguinal WAT and one plate of 10 cm for BAT that is isolated from five mice.
17. Plate the cells in a complete medium on collagen coated plates.

18. Incubate the cells for 1 to 2 h in 37°C, 5% CO_2 humid incubator (can extend up to 4 h). Discard the medium and replenish with new complete medium. (Important steps to remove other contaminant.)
19. Grow primary adipocytes to 95%–97% confluence in complete medium by replacing it with new complete medium every 2 days.

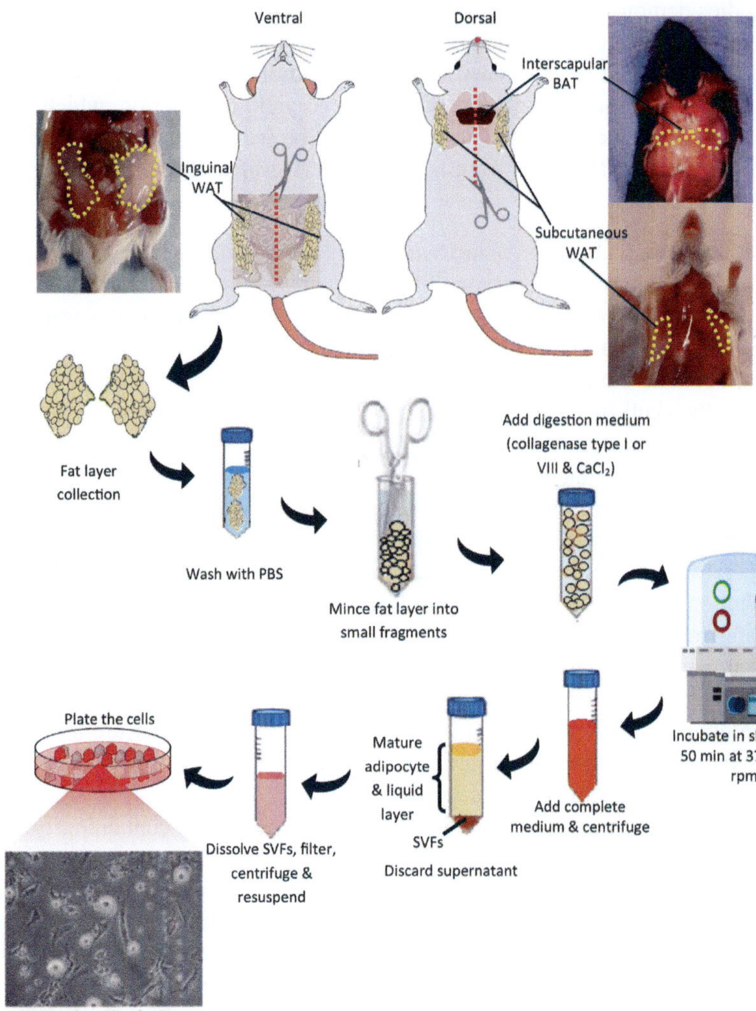

FIG. 3

Schematic overview of ADSCs isolation from rodent.

The image of microscopic ADSCs culture is adopted from Huang, S.J., Fu, R.H., Shyu, W.C., Liu, S.P., Jong, G.P., Chiu, Y.W., et al., 2013. Adipose-derived stem cells: isolation, characterization, and differentiation potential [research support, non-U.S. Gov't review]. Cell Transplant. 22 (4), 701–709. https://doi.org/10.3727/096368912X655127 and the image of mice for Interscapular BAT is adopted from Hofmann, S.M., et al., 2007. Adipocyte LDL receptor–related protein–1 expression modulates postprandial lipid transport and glucose homeostasis in mice. J. Clin. Invest. 117, 3271–3282. https://doi.org/10.1172/JCI31929.

3.7 Neural stem cell isolation

An adult brain was believed to not be able to generate new neurons. This belief was sustained until 1960s when a group of scientist confronts this canon. Neurogenesis occurs essentially in the subventricular zone (SVZ) at the forebrain of adult brain including the subgranular zone (SGZ) within the hippocampus. Several previous rodent studies demonstrated through in vitro and in vivo that SVZ progenitor cells have potential to act as neural stem cells. Although isolation of neural cells from postnatal or embryonic brain is easy, isolation of adult neural stem cells by using protease digestion technique will give an increased yield of isolation, which is not found in other available method. Thus, in this section we provide a detailed procedure to isolate neural stem cells from adult mouse (Deshpande et al., 2019; Volker et al., 2019; Wu et al., 2008).

1. Use a mouse aged between 5 and 8 months old. Brain tissue from one mouse is adequate to obtain neural stem cell line.
2. Euthanize the mouse by exposing to 100% CO_2 and proceed with cervical dislocation to ensure death (Fig. 4).
3. Decontaminate the mouse with 70% alcohol.
4. Expose the skull by cutting the skin along midline. Remove the hair and skin while harvesting brain tissue.
5. Cut the skull along the midline using thin, sharp scissors to expose the brain tissue. Scoop or push the brain tissue gently out of skull using scalpel and immediately let it fall into a petri dish containing DMEM/F-12 in room temperature (25°C). You may also do immediate rinse the brain with 70% alcohol once and then continue with washing twice using PBS before transferring into petri dish containing DMEM/F-12 as an extra precaution from contamination.
6. Bring the petri dish into laminar flow to proceed with next steps. Aseptic techniques are applied to minimize possible contamination. Ensure that the brain tissue is covered and hydrated with DMEM/F-12 all the time form this step forward during cell isolation process.
7. Transfer the brain tissue immediately into a new petri dish containing fresh DMEM/F-12 medium.
8. Prior to cell isolation, mince the brain tissue using a small sharp scissors and forceps into small pieces. Mince the tissue as finely as possible.
9. Remove the meninges by gentle cut to reduce contamination of endothelial cells.
10. Transfer the minced brain tissue with the DMEM/F-12 medium into a 50 mL conical tube using pipette and the pieces are allowed to settle to the bottom of the tube. Meanwhile, prepare the reagents required for the next steps.
11. From the papain dissociation kit, take the albumin-ovomucoid inhibitor mixture and add 32 mL of cold EBSS into it. Allow dissolution to take place after that reconstitute desirable amount to use and store the residue for future use. It is important to avoid formation of air bubbles during mixing.
12. Meanwhile, prepare another mixture. Use papain vial as provided in the dissociation kit to add 5 mL of EBSS solution. Incubate for 10 min at 37°C or

until the solution turns clear as the appearance of normal media indicating that the papain is fully dissolved. To dissolve the minced brain of one adult mouse, prepare two papain vials filled with 5 mL EBSS solution each.

13. After that, add 500 µL of EBSS to the DNase vial provided in the kit and mix gently, to avoid DNase degradation. (Prepare two DNase vial if you prepare two papain vial.) Add 250 µL of this solution to each papain vial prepared in step 12. This mixture will give rise to 20 U/mL papain and 0.005% DNase. The remaining DNase will be used in step 20.
14. From the 50 mL conical tube in step 10, carefully aspirate the medium of minced tissue without sucking up the minced tissue.
15. Add papain/DNase solution to the minced tissue immediately with the total volume of 7.5 to 10 mL/brain. Immediately close the lid, mix gently by ensuring no air bubbles formation. Use the solution from one or two papain vial prepared in steps 12 and 13.
16. Constantly agitate the 50 mL tube containing the tissue for 60 to 90 min using shaker in 37°C incubator. Manually swirl the mixture every 10 min and ensure that the mixture turns cloudy indicating that the cells are dissociated from minced tissue.
17. After that, by using a glass Pasteur pipette, triturate the mixture 20 to 30 times until the solution turns cloudy. Prevent from air bubbles formation. Since adult tissue is more fibrous and studier, it is necessary to prolong the trituration duration.
18. Filter the mixture into a new 50 mL conical tube using a cell strainer to get rid of any bigger clump. In this step, it is recommended to vigorously mix the solution to increase yield of the cells.
19. Centrifuge the strained mixture at 300 × g for 5 min in room temperature (25°C).
20. Meanwhile, prepare solutions to resuspend the cell pellet from step 19. In a new sterile 15 mL conical tube mix 300 µL albumin-ovomucoid inhibitor solution from step 11 with 2.7 mL EBSS prior to adding 150 µL of DNase solution that is kept aside at step 13. Reminder: above volume is prepared for only one papain vial. Kindly double all the volumes in this step if two papain vials were used for dissociation in step 15.
21. After centrifugation, aspirate and discard the supernatant and gently resuspend cell pellet immediately with the inhibitor solution prepared in step 20 while preventing from air bubble formation during mixing.
22. Prepare a discontinuous density gradient by adding 5 mL of albumin-ovomucoid inhibitor solution into a new 15 mL conical tube. After that overlay the cell suspension prepared in step 21 carefully on top of the inhibitor solution. Centrifuge at 400 × g for 5 min at room temperature (25°C). Reminder: use 50 mL tube and double the of inhibitor solution if the volumes are doubled as described in step 20.
23. Aspirate out the supernatant and resuspend the cell pellet with sterile NSCs medium supplemented with FGF (8 to 10 ng/mL) and EGF (20 ng/mL). Count the cells using 4% trypan blue and seed the cells accordingly in a nonadherent cell culture flask. Incubate at 37°C on a shaker. Maintain the plated cells for 3 to 4 days until cell clumps appear in suspension.

24. After 3 to 4 days, live stem/progenitor cell clumps in suspension will be visible under microscopic view.
25. Once cell clumps is observed, pipette the mixture up and down 5 to 10 times to dislodge the dead cells and debris from live stem/progenitor cell clumps. Wet the pipette with medium before use to avoid the cells from sticking to the pipette wall.
26. View under microscope to ensure dislodging prior to pipetting the solution into new sterile 50 mL conical tube.
27. Leave the tube to rest for 10 to 15 min to let the heavier cell clumps to settle to the bottom of the flask leaving the debris at supernatant.
28. Slowly aspirate out the supernatant containing debris and add 2 to 3 mL sterile NSC medium to the pelleted cells to avoid drying.
29. Collect the remaining cell clumps from the culture flask by washing with 5 to 6 mL of NSC medium and transfer into the 50 mL tube from step 26.
30. Repeat steps 27 and 28.
31. Centrifuge at 3000 × *g* for 15 min in room temperature (25°C) and aspirate out the supernatant. The supernatant can be saved to use as conditioned medium to passage the stem/progenitor cell clumps (neurospheres).
32. Add approximately 1 mL of accutase to the cell pellet depending on the pellet size. (Increase the Accutase volume as the pellet size increase.)
33. Incubate the cells for 4 to 5 min in 37°C incubator on a shaker.
34. Later, pipette the cells up and down several times to disperse the neurospheres into single cells.
35. Add NSC medium twice the accutase to deactivate it once sufficient smaller neurospheres is achieved.
36. Pipette the cells using 1000 μL micropipette tip for about 30 to 40 times by pushing the pipette tip to the tube bottom in order to dissociate the smaller size of spheres.
37. Triturate further if it is required after confirming dissociation under the microscope.
38. Centrifuge the cells at 500 × *g* for 1.5 min at room temperature (25°C) and resuspend in NSC medium. Proceed with cell counting and viability using trypan blue prior to seeding.
39. Prepare poly-ornithine or fibronectin-coated plates or dish 1–2 days before use.
40. Keep coated plate in room temperature (25°C) more than 1.5 h before use and wash the plates twice with distilled deionized H_2O prior to seeding.
41. Plate the whole neurospheres on the poly-ornithine/fibronectin-coated plates in NSC media supplemented with EGF (20 ng/mL) and FGF (8–10 ng/mL) to maintain neural stem cell multipotency. The characteristic of NSC will be maintained if the cells are not supplemented with lineage-specific growth factors.
42. Single cells obtained from neurospheres dissociation can also be seeded at density of $2–4 \times 10^5$ cells/cm^2.
43. Change the media every 3 to 4 days until further experiment is required.

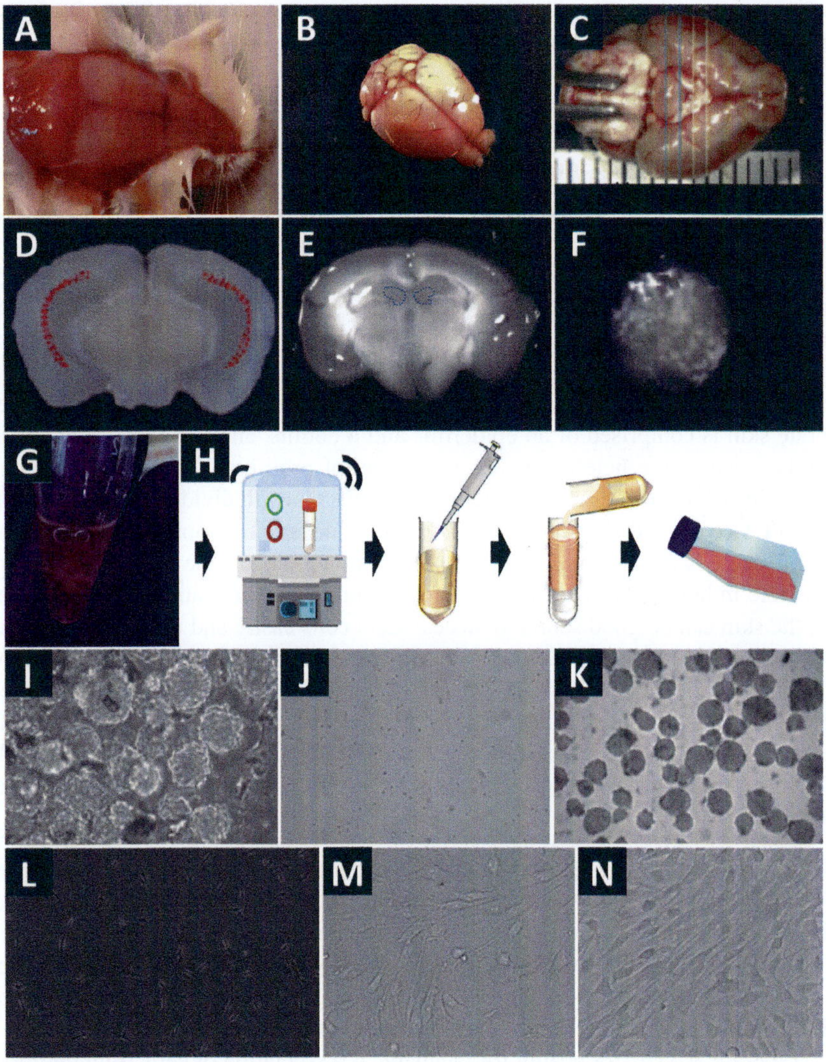

FIG. 4

Isolation of neural stem cells from mice. (A) Skin is cut open exposing the skull. (B) Isolate brain tissue from the skull by gentle scooping. (C) To extract specific region which are the dentate gyrus (DG) and subventricular zone (SVZ), the area between the blue lines and white lines have to be dissected out. (D) From the dissected region, the dashed red lines represent SVZ and (E) dashed blue lines represent DG. The region has to be separated gently and (F) minced into small fragments. In the case that the researchers don't require specific region of the brain, majority of the brain part can be minced by only removing the meninges. (G) Add minced tissue to the dissociation solution and (H) incubate in shaker incubator for 60–90 min at 37°C. Centrifuge, resuspend with inhibitor solution. Cells are added to density gradient solution and centrifuge. Resuspend pallet with complete medium and incubate until further steps required. (I–K) Phase contrast images of neurospheres captured on day 7 of postisolation, after dispersion of neurospheres into single cells, and after passage 1, respectively. (L–N) To obtain monolayer cells, culture the cells on coated flask. (L) Monolayer culture of neural precursor cells, (M) after 1 day of seeding and (N) after 3 days of culture.

Images are adopted from Kim, J.Y., Lee, J.H., Sun, W., 2016. Isolation and culture of adult neural stem cells from the mouse subcallosal zone. J. Vis. Exp. 188, 54929. doi:10.3791/54929, Bernas, S.N., Leiter, O., Walker, T.L., Kempermann, G., 2017. Isolation, culture and differentiation of adult hippocampal precursor cells. Bio Protocol. 7 (21). doi:10.21769/BioProtoc.2603, and Ahmed, K.M.A.A., Isaksen, T.J., Yamashita, T., 2021. Protocol for mouse adult neural stem cell isolation and culture. Cell Press 2 (2), 100522. https://doi.org/10.1016/j.xpro.2021.10052.

3.8 Stem cell isolation from epidermis

The skin is a protective barrier against foreign assaults and harmful elements. It is the largest organ of human body. The epidermis was found to have different unique compartments such as the interfollicular epidermis, hair follicles (HFs), sweat glands, and sebaceous glands (SGs) that house resident stem and progenitor cells, which play a major role to replenish itself. Together with other tissues, the epidermis also was found to have stem cells and due to its constant renewal ability and high propensity for repair, the skin becomes a tissue of choice to explore stem cell biology. The vertebrate skin is comprised of an epidermis and a dermis, and the basal layer of epidermis is where the heterogeneous population of proliferative and differentiating cells resides, known as epidermal stem (EPS) cells that will divide into transit amplifying (TA) cells and this TA cells give rise to different compartments epidermis layer. EPS cells are also important for tumor initiation and gene therapy in addition to its role in homeostasis and wound repair. Liang and Bickenbach (2002) suggest that the skin can be good source to access stem cells easily and thus reprogrammed into different cell lineage. Identifying, isolating, and culturing pure EPS cells have been challenging; thus, here we will discuss the enzyme treatments and culture selection techniques in order to isolate distinct stem cell population namely mesenchymal stem cells (MSCs), skin-derived precursors (SKPs), and epidermal stem cells (EpiSCs) from rodent dermis and epidermis (Forni et al., 2015; Kostic et al., 2017; Soteriou et al., 2016).

1. Euthanize 2- to 3-month-old male Balb/c mice with 100% CO_2 (slow fill rates of 20%–30% of the chamber volume/min).
2. Initiate tissue collection immediately after euthanasia. Shave the fur and hypodermis off the back skin of the mouse. Decontaminate the area of interest with 70% alcohol.
3. Cut the skin into 2 to 3 mm^2 fragments and incubate into HBSS containing Liberase at 37°C for 60 min for mechanical dissociation of cells from dermis to obtain MSCs and SKPs.
4. After that, add DMEM supplemented with 10% FBS double the amount of HBSS containing Liberase to deactivate the mechanical digestion. Triturate several times to mix well.
5. Aspirate the cell suspension and distribute equally into two separate 50 mL conical tube prior to centrifugation. Centrifuge for 10 min at 700 $\times g$. Resuspend the cells of two different tubes in SKP and mesenchymal culture media, respectively. Precede cell counting using trypan blue for both tubes prior to plating.
6. The remaining epidermal layers from step 5 are incubated into Dispase 100-caseinolytic units/mL at 37°C for 30 min. This is another cycle of digestion to obtain the third cell population, the EpiSCs.
7. Following mechanical dissociation, add DMEM supplemented with 10% FBS double the amount of dispase to deactivate the mechanical digestion. Triturate several times to mix well and centrifuge for 10 min at 700 $\times g$.

8. Discard the supernatant and resuspend the cells with EpiSC culture medium. Precede cell counting using trypan blue prior to plating.
9. Seed the cell in collagen IV coated plates. Twenty-four hours after plating, remove the medium, continue washing with HBSS, and then add 0.1% trypsin to the cells for 2 min at 37°C. The non-EpiSCs will be floating, aspirate the supernatant and then wash the plates with HBSS.
10. Replenish the EpiSC culture medium, continue incubation and replace the media 2–3 days once until cell becomes 70%–80% confluent and ready for first passage.
11. The EpiSCs is required to subcultured in collagen IV coated plates supplemented with KSF medium containing EPB, 20 ng/mL EGF, and 1% ampicillin-streptomycin.
12. Meanwhile, the MSCs were maintained in adherent flasks in DMEM supplemented with 10% FBS and 1% ampicillin-streptomycin.
13. Maintain the SKPs in DMEM/F-12 at a ratio of 3:1 containing 20 ng/mL EGF, 40 ng/mL bFGF, and 1% B-27 supplement.
14. Incubate the cells under controlled humidity at 37°C with 2% CO_2 and continue to subculture cells as per requirement of the experiment for further analysis.

3.9 Cryopreservation

1. Discard the cell culture medium from flask.
2. Wash the adherent cell layer with sterile calcium- and magnesium-free 1× PBS twice to remove debris and loosely attached cells.
3. Discard the PBS from the flask.
4. Add ~2 mL of warm 0.25% trypsin-EDTA solution and incubate in incubator for 3–5 min.
5. View under microscope to confirm the cell detachment. The flask may need a few firm gentle taps if the cells are not completely detached.
6. Once the cells are detached, add cell culture complete media supplemented with 5%–10% FBS to the flask (1:1 ratio to trypsin-EDTA).
7. Collect the cell suspension and transfer into a 15 mL or 50 mL tube and centrifuge at 200×*g* or 1200 rpm for 5 min.
8. Discard the supernatant with sterile pipette tip without disturbing the pellet.
9. Add 10 mL of cell culture medium to the pellet and resuspend evenly and then count the cells with trypan blue stain under the microscope.
10. After counting, centrifuge the cell suspension again, discard the supernatant, and resuspend the pellet quickly in freeze medium at concentration of 1×10^6 cells/mL. Aliquot 1 mL of cells into cryovial.
11. After aliquoting each vial with 1 mL of 1×10^6 cells, immediately seal the vial and place in freezing container containing 100% isopropyl alcohol at −80°C for 24 h. Later, transfer the vials to the vapor phase of liquid nitrogen tank for indefinite storage.

4. Conclusion

Rodent stem cell isolation and manipulation have led to various clinical findings and applications. It is a promising approach for future research in bringing this robust technology to clinical settings that benefits therapeutic medicine. In this regard, the procedures elucidated here will be an important guideline for scientists world-wide to move forward with a balanced moral and ethical concern.

Abbreviations

%	percentage
~	approximately
°C	degree Celsius
7-AAD	7-aminoactinomycin D
ADSCs	adipose-derived stem cells
APC	antibody-photosensitizer conjugate
ASCs	adult stem cells
BAT	brown adipose tissue
bFGF	basic fibroblast growth factor
BM	bone marrow
BMSCs	bone marrow derived stem cells
BPE	bovine pituitary extract
$CaCl_2$	calcium chloride
c-Kit	receptor tyrosine-protein kinase
cm	centimeter
cm^2	square centimeter
CO_2	carbon dioxide
d.p.c.	days post coitum
DMEM	Dulbecco's modified eagle medium
DMSO	dimethyl sulfoxide
EBSS	Earle's balanced salt solution
EDTA	ethylenediaminetetraacetic acid
EGF	epidermal growth factor
EpiSCs	epidermal stem cells
EPS	epidermal stem
ES	embryonic stem
ESCs	embryonic stem cells
FACS	fluorescence-activated cell sorting
FBS	fetal bovine serum
FCS	fetal calf serum
FGF	fibroblast growth factor
FITC	fluorescein isothiocyanate
FMO	fluorescent minus one
g	gram
G	gauge
GSK	glycogen synthase kinase 3 inhibitor

H_2O	water
HBSS	Hank's balanced salt solution
HFs	hair follicles
HSCs	hematopoietic stem cells
IACUC	Institutional Animal Care and Use Committee
ICMs	from inner cell masses
IL-3	interleukin 3
IL-6	interleukin 6
IMDM	Iscove's modified Dulbecco's medium
IU	international unit
KCl	potassium chloride
KH_2PO_4	monopotassium phosphate
KOSR	knockout serum replacement
KSF	keratinocyte serum free
KSOM	Embryomax potassium supplemented simplex optimized media
L	liter
LIF	leukemia inhibitory factor
Lin	lineage marker
LSK	Lin-c-Kit+Sca1+ cells
MEF	mouse embryonic fibroblast
MEK	mitogen-activated protein kinase
MEM	minimum essential medium
mESCs	mouse embryonic stem cells
mL	milliliter
mM	millimolar
mm	millimeter
mm^3	cubic millimeter
MNCs	mononucleated cells
MSC	mesenchymal stem cell
MSCs	mesenchymal stem cells
Na_2HPO_4	disodium hydrogen phosphate
NaCl	sodium chloride
$NaHCO_3$	sodium bicarbonate
ng	nanogram
NSC	neural stem cell
PBS	phosphate-buffered saline
PE	phycoerythrin
pH	potential of hydrogen
PSN	penicillin-streptomycin-neomycin solution
rpm	revolutions per minute
Sca-1	stem cells antigen-1
SCF	stem cell factor
SCs	stem cells
SGs	sebaceous glands
SGZ	subgranular zone
SKPs	skin-derived precursors
SVFs	stromal vascular fractions

SVZ	subventricular zone
TA	transit amplifying cells
U	unit
UC	umbilical cord
UC-SCs	umbilical cord derived stem cells
WAT	white adipose tissue
xg	relative centrifugal force
μg	microgram
μL	microliter
μm	micrometer

References

Astori, G., Soncin, S., Lo Cicero, V., Siclari, F., Surder, D., Turchetto, L., et al., 2010. Bone marrow derived stem cells in regenerative medicine as advanced therapy medicinal products. Am. J. Transl. Res. 2 (3), 285–295.

Deshpande, K., Saatian, B., Martirosian, V., Lin, M., Julian, A., Neman, J., 2019. Isolation of neural stem cells from whole brain tissues of adult mice. [research support, U.S. Gov't, non-P.H.S. research support, non-U.S. Gov't]. Curr. Protoc. Stem Cell Biol. 49 (1), e80. https://doi.org/10.1002/cpsc.80.

Ferraro, G.A., Mizuno, H., Pallua, N., 2016. Adipose stem cells: from bench to bedside. Stem Cells Int. 2016, 6484038. https://doi.org/10.1155/2016/6484038.

Forni, M.F., Ramos Maia Lobba, A., Pereira Ferreira, A.H., Sogayar, M.C., 2015. Simultaneous isolation of three different stem cell populations from murine skin [research support, non-U.S. Gov't]. PLoS One 10 (10), e0140143. https://doi.org/10.1371/journal.pone.0140143.

Fortier, L.A., Travis, A.J., 2011. Stem cells in veterinary medicine [research support, N.I.H., extramural research support, non-U.S. Gov't review]. Stem Cell Res Ther 2 (1), 9. https://doi.org/10.1186/scrt50.

Guasch, G., Fuchs, E., 2005. Mice in the world of stem cell biology [research support, N.I.H., extramural research support, non-U.S. Gov't review]. Nat. Genet. 37 (11), 1201–1206. https://doi.org/10.1038/ng1667.

Herzog, E.L., Chai, L., Krause, D.S., 2003. Plasticity of marrow-derived stem cells [research support, U.S. Gov't, P.H.S. review]. Blood 102 (10), 3483–3493. https://doi.org/10.1182/blood-2003-05-1664.

Huang, S.J., Fu, R.H., Shyu, W.C., Liu, S.P., Jong, G.P., Chiu, Y.W., et al., 2013. Adipose-derived stem cells: isolation, characterization, and differentiation potential [research support, non-U.S. Gov't review]. Cell Transplant. 22 (4), 701–709. https://doi.org/10.3727/096368912X655127.

Huang, S., Xu, L., Sun, Y., Wu, T., Wang, K., Li, G., 2015. An improved protocol for isolation and culture of mesenchymal stem cells from mouse bone marrow. J. Orthop. Translat. 3 (1), 26–33. https://doi.org/10.1016/j.jot.2014.07.005.

Jozefczuk, J., Drews, K., Adjaye, J., 2012. Preparation of mouse embryonic fibroblast cells suitable for culturing human embryonic and induced pluripotent stem cells [research support, non-U.S. Gov't video-audio media]. J. Vis. Exp. (64). https://doi.org/10.3791/3854.

Kostic, L., Sedov, E., Soteriou, D., Yosefzon, Y., Fuchs, Y., 2017. Isolation of stem cells and progenitors from mouse epidermis. Curr. Protoc. Stem Cell Biol. 41, 1C 20 21–21C 20 11. https://doi.org/10.1002/cpsc.26.

Kumar, B., Madabushi, S.S., 2018. Identification and isolation of mice and human hematopoietic stem cells. Methods Mol. Biol. 1842, 55–68. https://doi.org/10.1007/978-1-4939-8697-2_4.

Lee, B.K.-H., 2013. In: Lenka, B.A.N. (Ed.), Pluripotent Stem Cells: Conditions and Techniques for Mouse Embryonic Stem Cell Derivation and Culture D., https://doi.org/10.5772/45917.

Li, W.W., Wei, Y.H., Li, H., Lai, D.M., Lin, T.N., 2013. Isolation and characterization of a novel strain of mesenchymal stem cells from mouse umbilical cord: potential application in cell-based therapy [research support, non-U.S. Gov't]. PLoS One 8 (8), e74478. https://doi.org/10.1371/journal.pone.0074478.

Liang, L., Bickenbach, J.R., 2002. Somatic epidermal stem cells can produce multiple cell lineages during development. Stem Cells 20 (1), 21–31. https://doi.org/10.1634/stemcells.20-1-21.

Sakinah, S., Priya, S.P., Mok, P.L., Munisvaradass, R., Teh, S.W., Sun, Z., Alzahrani, B., Abu Bakar, F., Chee, H.Y., Awang Hamat, R., He, G., Xiong, C., Joseph, N., Tong, J.B., Wu, X., Maniam, M., Samrot, A.V., Higuchi, A., Kumar, S.S., 2021. Stem cell therapy in dengue virus-infected BALB/C mice improves hepatic injury. Front. Cell Dev. Biol. 9, 637270. https://doi.org/10.3389/fcell.2021.637270.

Soteriou, D., Kostic, L., Sedov, E., Yosefzon, Y., Steller, H., Fuchs, Y., 2016. Isolating hair follicle stem cells and epidermal keratinocytes from dorsal mouse skin [video-audio media]. J. Vis. Exp. (110). https://doi.org/10.3791/53931.

Vadakke-Madathil, S., Limaye, L.S., Kale, V.P., Chaudhry, H.W., 2019. Progenitor Cells: Methods and Protocols, Methods in Molecular Biology: Flow Cytometry and Cell Sorting Using Hematopoietic Progenitor Cells. vol. 2029 Springer Science+Business Media, LLC, part of Springer Nature.

Volker, J., Engert, J., Volker, C., Bieniussa, L., Schendzielorz, P., Hagen, R., Rak, K., 2019. Isolation and characterization of neural stem cells from the rat inferior colliculus. Stem Cells Int. 2019, 5831240. https://doi.org/10.1155/2019/5831240.

Wu, Y., Liu, Y., Chesnut, J.D., Rao, M.S., 2008. Isolation of neural stem and precursor cells from rodent tissue [research support, N.I.H., extramural research support, non-U.S. Gov't]. Methods Mol. Biol. 438, 39–53. https://doi.org/10.1007/978-1-59745-133-8_5.

Yeoh, J.S., Ausema, A., Wierenga, P., de Haan, G., van Os, R., 2007. Mobilized peripheral blood stem cells provide rapid reconstitution but impaired long-term engraftment in a mouse model [research support, non-U.S. Gov't]. Bone Marrow Transplant. 39 (7), 401–409. https://doi.org/10.1038/sj.bmt.1705601.

Zhang, B., Zhang, J., Shi, H., Mao, F., Wang, J., Yan, Y., et al., 2018. A novel method to isolate mesenchymal stem cells from mouse umbilical cord. Mol. Med. Rep. 17 (1), 861–869. https://doi.org/10.3892/mmr.2017.7950.

CHAPTER

Protocols in stem cell culture 3

Tong Jiabei[a], Thilakavathy Karuppiah[a,b], Sun Zhong[b], Akon Higuchi[c,d,e], and Suresh Kumar Subbiah[f]

[a]*Department of Biomedical Science, Faculty of Medicine and Health Sciences, Universiti Putra Malaysia, Serdang, Selangor, Malaysia,* [b]*Genetics & Regenerative Medicine (ReGEN) Research Group, Faculty of Medicine and Health Sciences, Universiti Putra Malaysia, Serdang, Selangor, Malaysia,* [c]*State Key Laboratory of Ophthalmology, Optometry and Visual Science, Eye Hospital, Wenzhou Medical University, Wenzhou, Zhejiang, China,* [d]*Department of Chemical and Materials Engineering, National Central University, Taoyuan, Taiwan,* [e]*Center for Emergent Matter Science, Riken, Wako, Saitama, Japan,* [f]*Centre for Materials Engineering and Regenerative Medicine, Bharath Institute of Higher Education and Research, Chennai, Tamil Nadu, India*

1. Literature review

1.1 Mesenchymal stem cells (MSCs)

Mesenchymal stem cells (MSCs) can exist in the bone marrow, umbilical cord blood, adipose tissue, skeletal muscle, and other tissues, and possess a strong capacity for self-renewal and multipotent differentiation. MSCs are spindle-shaped adherent cells when cultured under in vitro conditions. As cells with high proliferative capacity, MSCs can effectively expand in vitro by forming distinct colonies and positively expressing mesenchymal molecular markers. Moreover, the International Society for Cellular Therapy (ISCT) has established criteria for identifying MSCs in 2006, which included: (1) adherent cells in spindle shape; (2) cells with positive expressions of CD73, CD90, and CD105 and negative expressions of CD11b, CD19, CD34, CD45, CD79α, and HLA-DR; and (3) cells which are able to effectively differentiate into adipocytes, osteocytes, and chondrocytes (Stabler et al., 2015; Tm, 2019).

1.1.1 Bone marrow-derived MSCs (BMSCs)

The bone marrow-derived MSCs are bone marrow stromal cells found within the marrow stroma of mammals. They are multipotent subgroups that can effectively differentiate into various cell types such as the bone, cartilage, fat, nerve, and myoblast. They provide mechanical support for the hematopoietic stem cells (HSCs) in the bone marrow and promote hematopoiesis through the secretion of multiple growth factors like IL-6, IL-11, LIF, M-CSF, and SCF (Shi and Gronthos, 2013; Wu et al., 2016).

Stem Cell Laboratory Techniques. https://doi.org/10.1016/B978-0-12-823729-8.00006-3

The most common methods used to isolate MSCs include the whole bone marrow technique and density gradient centrifugation. For instance, the whole bone marrow uses the adhesive features of stem cells in a low serum medium and can facilitate the purification and amplification of MSCs by removing nonadherent cells through liquid periodical changes. Density gradient centrifugation refers to a technique where monocytes are isolated for adherent cultures based on the different specific gravities of other cells in the bone marrow. However, not much difference has been found in nature for both the methods discussed above. Flow cytometry and magnetic-activated cell sorting are also utilized for the isolation and purification of MSCs as scientists have better understanding of their various surface antigens. However, cells sorted by the latter two methods can proliferate slowly and have other issues. High costs and technical difficulties have also limited their widespread use (Li et al., 2017).

1.1.2 Human umbilical-cord-blood-derived mesenchymal stem cells (hUCB-MSCs)

Umbilical cords are considered medical waste after delivery but are available in large quantities and can be easily harvested as well as transported without ethical concerns. Many MSCs may be amplified from umbilical cords, which have been widely used as the seed source of MSCs. Human umbilical-cord-blood-derived mesenchymal stem cells (hUCB-MSCs), similar to the bone marrow MSCs and MSCs derived from other sources, have been found to be adherent and express different markers of stem cells including CD10, CD13, CD29, CD44, CD90, and CD105, rather than the markers related to hematopoiesis. Besides, hUCB-MSCs can be easily isolated as well as cultured and their applications are unhindered by ethical issues. Hence, they have been widely utilized in the clinical practice (George et al., 2018; Ross et al., 2019).

1.1.3 Adipose tissue-derived stem cells (ADSCs)

The matrix of human subcutaneous connective tissue or the aspirate of knee local adipose tissue contains stem cells whose phenotypes are similar to that of BMSCs. These cells are termed as adipose tissue-derived stem cells (ADSCs) and express nearly the same CD markers as BMSCs. However, ADSCs express specific CD49d (integrin $\alpha4$), which is not expressed in BMSCs, while BMSCs express CD106 (vascular cell adhesion molecule), but it is not expressed by ADSCs. Similar to BMSCs, a series of different factors can boost the potential of adult stem cells and may also aid to transform ADSCs into mature adipocytes, osteoblasts, chondrocytes, muscle cells, endothelial progenitor cells, and nerve cells (Lee et al., 2012; Park et al., 2015).

1.1.4 Corneal stem cells

The limbus of the cornea is located at the junction of the cornea, conjunctiva, and sclera. It differs from conjunctiva as it does not have goblet cells and is about 1 mm wide; the epithelial cell layer contains over 10 layers and is irregularly arranged. Moreover, the cells are rather small and cylindrical and possess nuclear hyperchromatism, while its deep basal cell consists of a layer composed of small cylindrical or cuboidal cells with the nucleus being ovoid and parallel to the surface. In addition, the epithelial structure forms a special "fence," containing pigment and a rich vascular network, which can effectively maintain contact with the basement membrane. Limbus cornea stem cells are a subgroup of limbus cornea cells that possess unique biological characteristics such as longer survival, low differentiation, long cell cycle, short DNA synthesis period, and high differentiation as well as proliferation potentials. Stem cells can divide to produce two different daughter cells; one maintains the phenotype of mother cell and can become a stem cell, while the other evolves into an epithelial cell with distinct cellular functions. Stem cells found in self-renewing tissues are thought to be the main origin of various cell lineages as well as the potential source of cell proliferation and differentiation (Dulak et al., 2015; Lachaud et al., 2019; Sasamoto et al., 2019).

1.2 Hematopoietic stem cells

Hematopoietic stem cells (HSCs) is another kind of pluripotent stem cell that exists in hematopoietic tissue. This kind of cell is primarily a primitive hematopoietic cell that can differentiate and produce various mature blood cells (red blood cells, macrophages, lymphocytes, etc.) to maintain functions of the immune system under physiological conditions. Throughout the life cycle, hematopoietic stem cells can regulated to produce all types of mature blood cells, including red blood cells, white blood cells, and platelets. HSCs exhibit two distinct biological characteristics: self-renewal and multilineage differentiation. Adult hematopoietic stem cells are usually stored in the bone marrow in a quiescent state; however, they can undergo self-renewal or differentiation only when activated. Therefore, HSCs transplantation is important in the treatment of several hematopoietic diseases such as leukemia, aplastic anemia, lymphoma, multiple myeloma, and immunodeficiency. Moreover, the hematopoietic microenvironment can also play a pivotal role in the growth and development of HSCs and hematopoietic regulation. Therefore, HSCs and hematopoietic microenvironment play a key part in the recovery of the damaged hematopoietic system functions (Economescu and Rubach, 2015; Schiesser, 2019; Klump, 2020).

Human HSCs first appears in the yolk sac at an embryonic age of 2–3 weeks. During early embryonic period (around 2 to 3 months), HSCs can migrate to the liver and spleen, and then, it can translocate into the bone marrow in the 5th month. Bone marrow is the main source of HSCs from the end of the embryo to the birth. HSCs can be divided into proliferative and static state and are constantly switching between these two states. More than 95% of HSCs exist in the static state of G0 and do not carry out DNA synthesis as well as mitosis. It has been found that only if the body is subjected to stress conditions such as hypoxia, infections, blood loss, or trauma, the hematopoietic stem cells will enter into the proliferation state from the resting state to meet the demand for stimulating hematopoiesis. HSCs can not only maintain the stability of its quantity through self-renewal but also can function as directional progenitor cells of each line, which then can effectively differentiate and develop into naive as well as mature terminally differentiated blood cells (Schiesser, 2019).

1.3 Biomaterials

Human adult stem cells such as adipose-derived stem cells and bone marrow stem cells can be cultured on conventional polystyrene tissue culture polystyrene dishes. However, we should develop sophisticated and designed cell culture biomaterials, when we would like to develop biomaterials where stem cells are cultured and harvested without usage of conventional enzyme digestion method by external stimuli of (a) temperature change of cell culture biomaterials or (b) light irradiation into cell culture biomaterials. In this section, we discuss these sophisticated and designed smart (intelligent) biomaterials for stem cell cultivation.

1.3.1 Stem cells cultivation and harvest on photo-responsive biomaterials

A variety of methods have been proposed for changing the surface characteristics of cell culture biomaterials, which regulate surface hydrophilicity-hydrophobicity (Weihua et al., 2009; Brun et al., 2010; Nandivada et al., 2010; Higuchi et al., 2014; Yang et al., 2020). Especially, the light irradiation as the trigger for the phase change (transition) of stimulus-responsive biomaterials is valuable because light irradiation is easily and spatially controlled on biomaterial surface more accurately than other external stimuli. Azobenzene or spiropyran was typically selected as the key molecules for reversible photo-responsive sites in photo-responsive cell culture biomaterials.

Light-induced *E* (*trans*)-*Z* (*cis*) isomerization can be generated by azobenzene, which is extensively fast (within 20 ps) and this light-induced isomerization is known to be reversible (Auernheimer et al., 2005). Auernheimer and his colleagues developed PMMA (polymethylmethacrylate) surface immobilized with cRGDfK (cyclic Arg-Gly-Asp-Phe-Lys) peptide ($\alpha_V\beta_3$ and $\alpha_V\beta_5$ integrin binding site), which were bound to joint segments and azobenzene molecules (Auernheimer et al., 2005). This study showed photo-switched cell attachment on the surface. Fig. 1A describes the schematic illustration of cell attachment and detachment on the interface immobilized with molecules including light-twistable azobenzene and cRGDfK peptide (Delaittre et al., 2012). When UV light at 366 nm was irradiated on the interface, which induces $E \rightarrow Z$ isomerization of azobenzene, mouse fibroblasts, MC3T3, was found to be detached with hiding of cell attachment site because of close distance of cyclic RGD (Arg-Gly-Asp) peptide to the interface (Fig. 1A) (Delaittre et al., 2012). Cell attachment rate was shifted from 40% to 20% when $E \rightarrow Z$ isomerization was generated on this interface (Auernheimer et al., 2005). Relatively small effect of switchable regulation of cell adhesion and release by light irradiation in this work was found to be because of nonspecific cell attachment on base interface made of PMMA. Therefore, Liu and his colleagues designed to utilize PEG (polyethylene glycol) as a linker of molecules that contain GRGDS (Gly-Arg-Gly-Asp-Ser)

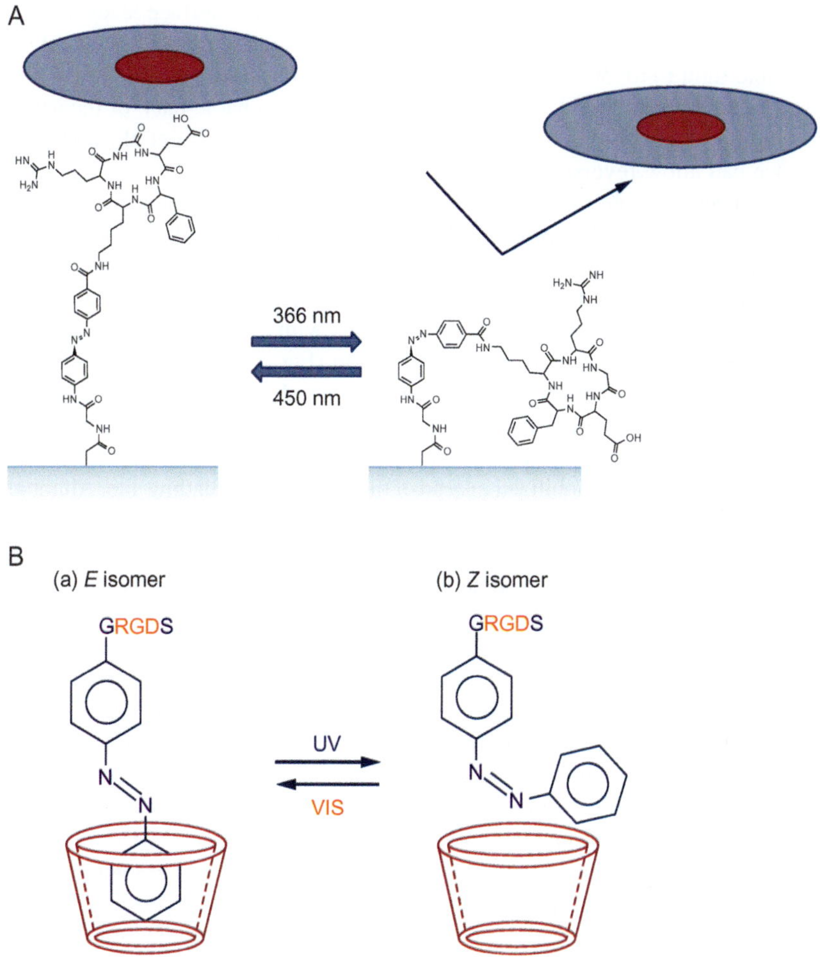

FIG. 1

Photoresponsive polymer interfaces. (A) The illustration of cell attachment and release on the interface having nanosegments binding light-twistable azobenzene linker and cRGDfK (Delaittre et al., 2012). (B) Reversible inclusion of azobenzene by α-cyclodextrin controlled by light irradiation. Azobenzene generates an inclusion complex with α-cyclodextrin by host-guest recognition and changes from the stable *E* form to the unstable *Z* form by irradiation of UV light, then reverts to the *E* form by irradiation of VIS (Gong et al., 2013).

oligopeptide and azobenzene on PEG-immobilized SAM (self-assembled monolayer) (Liu et al., 2009). PEG can work to reduce nonspecific adhesion of proteins that subsequently can reduce nonspecific cell attachment. On irradiation of UV light at 340–380 nm, the mice fibroblasts were able to be extensively detached from SAM interface, prepared from 99.9% PEG alkanethiolates and 0.1% of alkanethiolates containing azobenzene and GRGDS. PEG masking on the interface leads to protect nonspecific cell adsorption on the interface. After the releasement of the fibroblasts, the visible light (440–500 nm) was irradiated on the interface for 60 min, the fibroblast attachment was found, which is caused by the switchable back of the azobenzene configuration from *Z* form to *E* form on the interface (Liu et al., 2009). This interface can allow reversible regulation of cell attachment.

Azobenzene is able to be an inclusion molecule with cyclodextrin using host-guest recognition, which transforms from the thermodynamically stable *E* form to the unstable *Z* form by UV irradiation, and subsequently back to the *E* form by irradiation with visible light (VIS) (Fig. 1B) (Gong et al., 2013). The *E* form of azobenzene molecule can generate an inclusion complex with cyclodextrin and the *Z* form of azobenzene can be released from cyclodextrin (Fig. 1B) (Gong et al., 2013).

When we consider cultivation of human induced pluripotent stem cells (hiPSCs) and human embryonic stem cells (hESCs) on the micropatterned cyclodextrin surface, oligopeptides, which are necessary for keeping their pluripotent state and attachment on the interface, must be immobilized instead of the simple RGD binding site. Although it should be difficult to proceed mass production of hiPSCs and hESCs on micropatterned cyclodextrin interface, the reversible adhesion and detachment of hiPSCs and hESCs on a photo-responsive interface should be an interesting scientific topics.

Spiropyran can be utilized as a photo-responsive site. Nonionic spiropyran reversibly shifts into a zwitterionic merocyanine form by UV irradiation, which can be back to its original form by VIS irradiation (Higuchi et al., 2004). Higuchi and his colleagues reported a spiropyran-immobilized interface where spiropyran was utilized as the photo-responsive site to lead cell detachment by UV irradiation (Higuchi et al., 2004). Photo-induced releasement of mesenchymal stem (MS) cells was found from interface immobilized with a copolymer of methyl methacrylate (MMA), nitrobenzospiropyran (NSP), and poly(MMA-*co*-NSP) (Fig. 2A) (Higuchi et al., 2004). The hydrophilicity of the cell cultivation plates was reversibly changed when closed spiropyran (nonpolar structure) was induced to shift into the zwitterionic merocyanine form (polar structure) upon UV irradiation. Irradiation of UV on the photosensitive interface of poly(MMA-*co*-NSP)-immobilized plates reduced contact angles of water (10 degrees) and enhanced the radius of a water drop compared to those on the unexposed interface (Fig. 2B) (Higuchi et al., 2004). Photo-induced release of MS (KUSA-A1) cells, platelets, and fibrinogen from poly(MMA-*co*-NSP)-immobilized interface was found. On the other hand, no photo-induced detachment of fibrinogen, KUSA-A1 cells, and platelets from PMMA-immobilized dishes was found (Fig. 2C) (Higuchi et al., 2004). Furthermore, they succeeded to detach MS cells from a micro-patterned interface immobilized with poly(MMA-*co*-NSP), while MS cells did not release from a nonpoly(MMA-*co*-NSP)-immobilized domain, which resulted in micropatterned cell releasement of MS cells from poly(MMA-*co*-NSP)-immobilized dishes (Higuchi et al., 2004). It should be noted that the spatial resolution is high by patterned light irradiation; poly(MMA-*co*-NSP)-immobilized dishes are optimal for performing micropatterned detachment of adult stem cells, hiPSCs, and hESCs, or cells that were induced to differentiate from them in the future research.

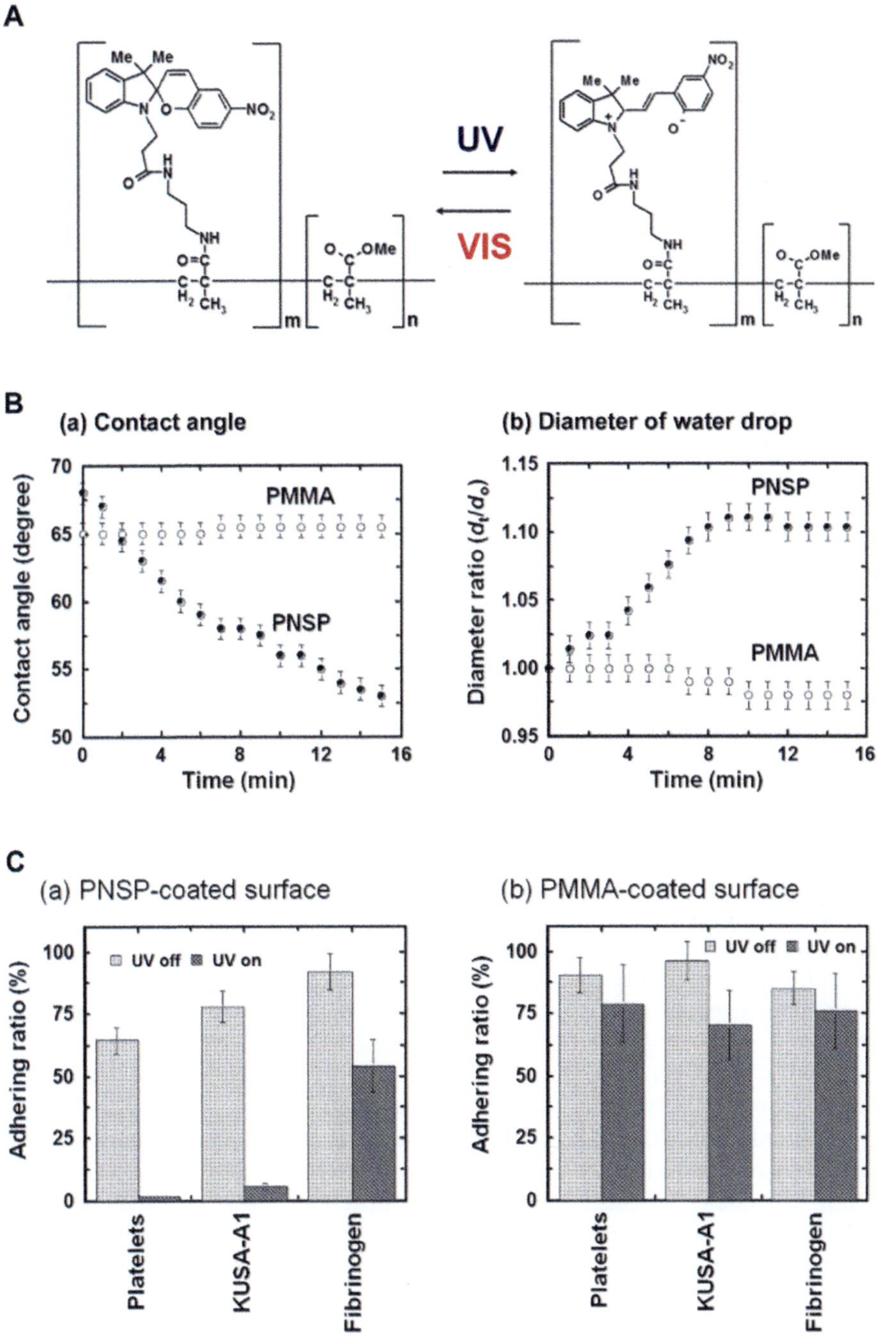

FIG. 2

Photoresponsive poly(MMA-*co*-NSP)-immobilized plates. (A) Illustration of the configurational change of poly(MMA-*co*-NSP) by irradiation of UV light. (B) Changes in hydrophilicity-hydrophobicity in poly(MMA-*co*-NSP)-immobilized and PMMA-immobilized dishes induced by irradiation of UV light. (a) Time dependence of the contact angle of water on poly(MMA-*co*-NSP)-immobilized and PMMA-immobilized dishes. (b) Time dependence of the ratio of the radius of water drops on poly(MMA-*co*-NSP) and PMMA-immobilized dishes. PNSP describes poly(MMA-*co*-NSP)-immobilized plates. (C) Light-induced release of fibrinogen, MS (KUSA-A1) cells, and platelets. (a) Fibrinogen, MS cells, and platelets on poly(MMA-*co*-NSP)-immobilized dishes. (b) Fibrinogen, MS cells, and platelets on PMMA-immobilized dishes. PNSP describes poly(MMA-*co*-NSP)-immobilized plates (Higuchi et al., 2004).

The biomaterial interface immobilized with segments made of cell binding site, which was protected with conjugation of photocleavable linker or a photo-cleavable group (cage), describes irreversible response to light irradiation. However, reversible regulation of cell adhesion and releasement on the biomaterial interface would be more interesting and valuable from scientific points. Kaneko and his colleagues reported a protocol for the regulation of cell attachment on amino-bearing interface by reversible binding of an antifouling segment, PEG, with a photocleavable linker (Kaneko et al., 2011). The photocleavable linker includes succinimidyl carbonate and alkyne at each end that were bound with photocleavable 2-nitrobenzyl ester. The molecule was bound between amines and azides, and the linkage could make a cleavage by near-UV light irradiation. HeLa cell attachment and protein adhesion could be reduced before light irradiation, whereas cell attachment and protein adhesion extensively enhanced by light irradiation that could be utilized for cellular patterning of striped lines with approximately 90 μm width (Kaneko et al., 2011). This technique should be interesting for micropatterning of stem cells on the biomaterial interface.

Photo-responsive interface should be adequate for the generation of micropatterned cell cultivation interfaces, which are not easy to make utilizing thermoresponsive polymer interface. The micropatterned cell cultivation interface should be powerful for cultivation of hiPSCs and hESCs, wherein (a) the hESCs or hiPSCs and (b) the feeder-layer cells could be separately cultivated on micropatterned interfaces in the same plates.

1.3.2 Stem cell cultivation and harvest on thermoresponsive biomaterials

Thermoresponsive interface composed of thermoresponsive polymeric materials with LCST (lower critical solution temperature) is interesting candidates for cell cultivation biomaterials because cells are released from the biomaterial interface without using an enzymatic digestion process and, instead, by reducing the temperature, e.g., to 18°C, which generates cell sheets or cell aggregates to be harvested. Poly(*N*-isopropyl acrylamide), PNIPA (Matsuura and Okano, 2013; Peng et al., 2015; Ekerdt et al., 2018; Kim et al., 2019; Sunga et al., 2019), is the most-studied thermoresponsive polymer, which is grafted on a plate interface to make cell release and to generate cell sheets by reducing the temperature. However, there are some other thermoresponsive polymeric biomaterials, which can be utilized for these objects. Elastin-like polypeptide, poloxamer (pluronic), methylcellulose (MC), hydroxybutyl chitosan, and xyloglucan are considered to use as thermoresponsive biomolecules, not only PNIPA (Higuchi et al., 2014).

1.3.2.1 Thermoresponsive polysaccharide interface

Polysaccharides such as MC, hydroxybutyl chitosan, and xyloglucan are able to work as thermoresponsive molecules when the balance of hydrophilic and hydrophobic sites in their molecular configurations is adequate.

MC hydrogels has been utilized as thermoresponsive materials to harvest cells or to create cell sheets by reducing the temperature of MC hydrogels where stem cells were cultivated (Huang et al., 2010).

Sung and his colleagues prepared three-dimensional (3-D) tissue constructs that contain uniformly distributed embryoid body (EB)-derived cells for usage as cardiomyocyte patches that employ a cell-sheet cultivation system with usage of MC as thermoresponsive hydrogels (Huang et al., 2010). Mouse ESCs (mESCs)-derived EBs having 200–300 μm in diameter were inoculated onto MC hydrogels (Huang et al., 2010). This MC has an optimal degree of methyl substitution and has thermoresponsive characteristics, which owns an adequate balance of hydrophilic and hydrophobic sites. MC in PBS (phosphate-buffered saline) shows a sol-gel transition around 30°C and, therefore, can be possible to be inserted into 96-well plates at 10–25°C, which form hydrogels in a cell culture incubator at 37°C (Huang et al., 2010).

Cell sheets of EB-derived cells were generated on the MC hydrogels. Beating cells were found in several domains of the cell sheets in this research (Huang et al., 2010). The sheets of EB-derived cells could spontaneously detach from the interface of the MC hydrogels when the plates were located at 20°C, and the cells were still beating (Huang et al., 2010). Then, the EB-derived cell sheets were embedded between slices of porous scaffolds (pore size = 140 μm, and 90% porosity). The scaffold sandwiches were then sutured and cultivated for 28 days (Huang et al., 2010). Unfortunately, visible contraction of the scaffold sandwiches was not obtained in this research, although beating cells were found in isolated sites of the scaffold sandwiches. It seems that the electromechanical integration among the beating areas was incomplete because of the EB-derived cardiac cells with low purity. Therefore, the generated force was not sufficient to generate the scaffold sandwiches to contract (Huang et al., 2010). However, this work indicated the possible potential of fabricating 3-D tissue constructs, which were made of cell sheets and sliced scaffolds for the treatment of the damaged tissue or organ, although it is important to study how to create cell sheets made cardiac cells derived from hESCs, hiPSCs, or MS cells with high purity.

1.3.2.2 Synthetic thermoresponsive polymer interface

Pluronics (poloxamer) that contain PEO (polyethylene oxide)-PPO (polypropylene oxide)-PEO triblocks has amphiphilic characteristics. Pluronics form lyotropic liquid crystal hydrogels because of their hydrophobic PPO segments and hydrophilic PEO segments (Higuchi et al., 2005). Pluronics show a phase transition temperature around 29°C, which is almost the same to the LCST of PNIPA (Higuchi et al., 2006) when PPO and PEO with adequate molecular weights are utilized. Higuchi and his colleagues generated pluronic-grafted plates by coating carbonyldiimidazole-activated pluronics on poly-D-lysine-immobilized plates (Higuchi et al., 2006), and cells were extensively released from both pluronic F68-grafted plates and pluronic F127-grafted plates by reducing the temperature of the plates into 10°C (Higuchi et al., 2006). The cell release can be generated by the temperature-dependent dehydration and hydration characteristics of pluronics, which are similar phenomena to

PNIPA-grafted plates. Pluronics exhibit bioinert properties. Then, the proliferation and differentiation of hESCs and hiPSCs in pluronic hydrogels and on pluronic-immobilized plates should be an interesting project for scientific research.

Zhang and his colleagues performed screening of thermoresponsive synthetic polymers using arrays consisting of 600 different polymers that were prepared from mixtures of 18 monomers in 7 different ratios using inkjet printing method for hESC (H9 and RH1) cultivation (Zhang et al., 2013). They reported that a family of thermoresponsive hydrogels made of 2-(acryloyloxyethyl) trimethylammonium chloride (AEtMA-Cl) and 2-(diethylamino)ethyl acrylate (DEAEA), DEAEA/AEtMA-Cl, maintained the long-term growth and pluripotent state of hESCs in defined media for more than 20 passages (Zhang et al., 2013). The hydrogels allowed enzyme-free cell passaging by changing the temperature of the hydrogels from 37°C to 14°C for 0.5h (Zhang et al., 2013). The chemically defined biomaterials prepared in this research will contribute to efficient and safe culture system of MS cells, hESCs, and hiPSCs for clinical application as well as research usage.

1.3.2.3 Thermoresponsive poly(*N*-isopropyl acrylamide) interface

PNIPA is extensively investigated as a thermoresponsive polymer. This is because PNIPA has sharp phase transition around 30°C that is similar to physiological temperature. PNIPA is consisted of repeating units of hydrophobic isopropyl moieties and hydrophilic amide moieties (Brun et al., 2010). In an aqueous solution, PNIPA chains show a reversible configurational transition from compact globules to expanded coils as the temperature decreased less than LCST of PNIPA (Brun et al., 2010). This configurational transition of PNIPA chains was used in the application of thermoresponsive materials for the cultivation and differentiation of ESCs and iPSCs, especially for releasing the cells from thermoresponsive PNIPA interface by lowering the temperature (Matsuura et al., 2011, 2012a; Kawamura et al., 2012; Masumoto et al., 2012). Okano and other investigators investigated the recovery of intact cell sheets prepared on PNIPA-grafted plates, which show changes of hydrophobic/hydrophilic properties of the plates in response to the change of temperature (Matsuura et al., 2011, 2012a,b; Kawamura et al., 2012; Masumoto et al., 2012).

PNIPA-grafted plates can allow the attachment of typical tissue cells. However, the attachment mechanism of the cells on PNIPA interface relies on a nonspecific interaction. It is required to add the cell-attachment site on the cell cultivation plates if less attached cells are cultivated on the PNIPA-grafted plates. Therefore, PNIPA-grafted plates immobilized the cell-binding site of RGDS (Arg-Gly-Asp-Ser) were also reported (Ebara et al., 2004a; Matsuura et al., 2012b).

Using electron beam irradiation, 2-carboxyisopropylacrylamide and *N*-isopropylacrylamide were copolymerized on the plates, and then, temperature-responsive poly(2-carboxyisopropylacrylamide-*co*-*N*-isopropylacrylamide), P(NIPA-*co*-CIPA), was immobilized on the plates. The P(NIPA-*co*-CIPA) includes carboxylic acid that leads to the attachment of RGDS utilizing water soluble cardodiimide in aqueous solution (Ebara et al., 2004a; Matsuura et al., 2012b). The interface allowed the spreading of HUVECs (Human Umbilical Vein Endothelial Cells) at 37°C and showed the release of HUVECs from the interface at 20°C (Ebara et al., 2004b). The "on-off" regulation of integrin-RGDS interaction by temperature control is valuable for cell property control because integrins on cell surfaces can interact with cell adhesion sites in ECMs and subsequently generate the activities of cell signal transduction.

Currently, the cell sheets are used to cure several types of tissue damages in a clinical application, such as periodontal, esophageal, cartilaginous, corneal, and heart tissue damages (Higuchi et al., 2014). Cell sheets generated from human stem cells were also prepared on PNIPA-grafted plates (Matsuura et al., 2012b; Higuchi et al., 2014).

Matsuura and his colleagues prepared cell sheets utilizing hiPSC-derived cardiac cells, which were cultivated in 3-D stirred suspension systems (Matsuura et al., 2012b). In this research, they induced differentiation of hiPSCs into EBs and cultivated EBs on thermoresponsive PNIPA-grafted plates. hiPSC-derived cardiac cells were generated at 1×10^8 cells/100 mL in the bioreactor cultivation vessel after 2 weeks, and cardiac troponin T (cardiac marker) was expressed around 79% (Matsuura et al., 2012b). The cells expressed synchronous and spontaneous beating, even after sheets of hiPSC-derived cardiac cells were released from PNIPA-immobilized plates. The extracellular action potential was extended between two sheets of hiPSC-derived cardiac cells among the cell sheets, which were overlaid (Matsuura et al., 2012b). These results indicate that the cell sheets of cardiomyocytes developed by hiPSC-derived cardiac cells on PNIPA-grafted plates can create multilayered cardiac cell sheets as thickened tissue of cardiomyocytes. The cell sheets thickness prepared from more than 15 cell sheets is a difficult subject currently (Sakaguchi et al., 2013; Higuchi et al., 2014) because of long preparation time and the luck of oxygen and nutrients in the multilayered cell sheets.

2. MSC-isolation of prime cells, cell culture and identification

2.1 Bone marrow-derived MSCs

2.1.1 Isolation of primary human BMSCs

Reagent:

DMEM/F12 medium, PBS, FBS, trypsin-EDTA (0.25%), 100 × penicillin-streptomycin (5000 U/mL).

Materials:

Nineteen gauge needles/60 mL syringes, T25 flasks, serological pipettes (5, 10, 25 mL), 1.5 mL centrifuge tubes, 15/50centrifuge mL tubes, pipet tips (10, 200, 1000 μL).

Equipment:

Centrifuge; biosafety cabinet; water bath; refrigerator (4°C and −20°C).

Procedure:

(1) Preparation of solution and reagent:

- **(i)** FBS: Make it melt at 4°C, divide into 50 mL tube, 45 mL into every tube, store at −20°C, one tube at 4°C for use.
- **(ii)** Culture medium (1 × penicillin-streptomycin, 20% FBS DMEM): 500 mL:100 mL FBS and 5 mL 100 × penicillin-streptomycin (5000 U/mL); add into 395 mL DMEM, mix, and store at 4°C.

(2) Add 2 mL of 4% sodium citrate into 50 mL tube.

(3) Add 10 mL human bone marrow into 50 mL tube, invert to mix gently.

(4) Dilute the human bone marrow by PBS (1:1, v/v): add 12 mL of PBS into 50 mL tube.

(5) Prepare 50 mL tubes with 15 mL Percoll into it, add diluted marrow slowly on the top of Percoll (blood must be added slowly, keep blood on the top of Percoll).

(6) Centrifuge at 400 × *g* for 20 min at room temperature with the accelerator and BRAKE OFF.

(7) After centrifuge, carefully collect MNC layer using new plastic transfer pipette into a new 50 mL tube.

(8) Dilute the MNC suspension (from step 7) into 15–20 mL per tube in 50 mL falcon tube (total 4 tubes) and dilute 1:1 in PBS; centrifuge at room temperature for 20 min at 300 × *g* with accelerator and BRAKE ON (5).

(9) Remove supernatant and re-suspend cell pellets by adding 1 mL supplemented DMEM (20% FBS) to each pellet and pipetting up and down repeatedly.

(10) Add 9 mL DMEM (20% FBS) into the tube and pipetting up and down repeatedly.

(11) Count and seed the cells by the density of 2×10^5 cells/cm^2.

2.1.2 Cell culture and freezing

Reagent:

Culture medium (1 × penicillin-streptomycin/20% FBS DMEM/F12).
Freezing medium: FBS:DMSO = 9:1.
Trypsin.
PBS.

Materials:

T25 flasks, serological pipettes (5, 10, 25 mL), 1.5 mL centrifuge tubes, 15/50 mL centrifuge tubes, pipet tips (10, 200, 1000 μL), freezing tube, liquid nitrogen tank, cell cryopreservation box.

Equipment:

Liquid nitrogen container, Refrigerator (4°C and −20°C).

Procedure:

(1) Culture and passage:

- **(i)** Add half amount of culture medium after primary cells have been planked for 24–48 h.

(ii) After adding culture medium 36–48 h, change medium and observe cells using microscope. When the cell confluence reaches 80%, detach cells by trypsin and then digest and passage cells. Plank the cell in the flask with density of 0.25–0.5 × 10^6 cells/T25.

(2) Cell freezing:

(i) Freeze the cells using freezing medium of FBS:DMSO = 9:1.

(ii) Detach cells by trypsin and neutralize by culture medium with the same amount. Centrifuge by 400 × *g*/5 min, re-suspend cell by 5 mL culture medium, and count the cell. Re-suspend the cells using freezing medium at the cell concentration of 5 × 10^5 cells/mL; every freezing tube contains 1–2 mL.

2.2 Umbilical cord blood MSCs

2.2.1 Isolation of primary human umbilical-cord-blood-derived mesenchymal stem cells (hUCB-MSCs)

Reagent:

Percoll (1.073 g/mL), 4% sodium citrate, DMEM, PBS, FBS, trypsin-EDTA (0.25%), 100× penicillin-streptomycin (5000 U/mL).

Materials:

T25 flask, serological pipettes (5, 10, 25 mL), 1.5 mL centrifuge tubes, 15/50 mL centrifuge tubes, pipette tips (10, 200, 1000 μL).

Equipment:

Centrifuge; biosafety cabinet; water bath; refrigerator (4°C and −20°C).

Procedure:

(1) Preparation of solution and reagent:

(i) FBS: Make it melt at 4°C, divide into 50 mL tube, 45 mL into every tube, store at −20°C, one tube at 4°C for use.

(ii) Culture medium (1 × penicillin-streptomycin/20% FBS DMEM): 500 mL:100 mL FBS and 5 mL 100 × penicillin-streptomycin; add into 395 mL DMEM, mix, and store at 4°C.

(2) Add 2 mL of 4% sodium citrate to 50 mL tube.

(3) Add 10 mL umbilical cord blood to 50 mL tube, invert to mix gently.

(4) Dilute the umbilical cord blood by PBS (1:1, v/v): add 12 mL of PBS to 50 mL tube.

(5) Prepare 50 mL tubes with 15 mL Percoll into it, add diluted blood slowly on the top of Percoll (blood must be added slowly, keep blood on the top of Percoll).

(6) Centrifuge at 400 × *g* for 20 min at room temperature with the accelerator and BRAKE OFF.

(7) After centrifuge, carefully collect MNC layer using new plastic transfer pipette into a new 50 mL tube.

(8) Dilute the MNC suspension (from step 7) into 15–20 mL per tube into 50 mL falcon tubes (4 tubes) and dilute 1:1 in PBS; centrifuge at room temperature for 20 min at 300 × *g* with accelerator and BRAKE ON (5).

(9) Remove supernatant and re-suspend cell pellets by adding 1 mL supplemented DMEM (20% FBS) to each pellet and pipetting up and down repeatedly.
(10) Add 9 mL DMEM (20% FBS) into the tube and pipetting up and down repeatedly.
(11) Count and seed the cells with the density of 2×10^5 cells/cm^2. Culture at 37°C, 5% CO_2.

2.2.2 Cell culture and freezing

Reagent:

Culture medium (1 × penicillin-streptomycin/20%FBS DMEM/F12), Freezing medium: FBS:DMSO = 9:1, Trypsin, PSB

Materials:

T25 flasks, serological pipettes (5, 10, 25 mL), 1.5 mL centrifuge tubes, 15/50 mL centrifuge tubes, pipette tips (10, 200, 1000 μL), freezing tubes, liquid nitrogen, cell cryopreservation box.

Equipment:

Liquid nitrogen container, Refrigerator (4°C and −20°C).

Procedure:

(1) Culture and passage:
  (i) Add half amount of culture medium after primary cells have been planked at 24–48 h.
  (ii) After adding culture medium at 36–48 h, change medium and observe cells by microscope. When the cell confluence reaches 80%, detach cells by trypsin and then digest and passage cells. Plank the cell in the flask with the density of $0.25–0.5 \times 10^6$ cells/T25.
(2) Cell freezing:
  (i) Freeze the cells by the freezing medium of FBS:DMSO = 9:1.
  (ii) Detach cells by trypsin and neutralize by culture medium with the same amount. Centrifuge by $400 \times g$/5 min, re-suspend cell by 5 mL culture medium, and count the cell. Re-suspend the cells by freezing medium at the cell concentration of 5×10^5 cells/mL; every freezing tube contains 1–2 mL.

2.3 Adipose tissue-derived stem cells

2.3.1 Isolation of primary ADSCs

Reagent:

DMEM/F12 medium, PBS, FBS, trypsin-EDTA (0.5%), collagenase type I dilution (0.1%), 100 × penicillin-streptomycin (5000 U/mL), 200 mM GlutaMAX.

Materials:

70 μm cell strainers, 100 mm cell culture dishes, T25 flasks, serological pipettes (5, 10, 25 mL), 1.5 mL centrifuge tubes, 15/50 mL centrifuge tubes, pipette tips (10, 200, 1000 μL).

Surgical tool:

Ophthalmic forceps, ophthalmic scissors, forceps (20–35 cm), tissue scissors (straight head, cured head), scalpels.

Equipment:

Centrifuge; biosafety cabinet; water bath; refrigerator (4°C and −20°C).

Procedure:

(1) Preparation of solution and reagent:
   - **(i)** FBS: Make it melt at 4°C, divide into 50 mL tube, 45 mL every tube, and store at −20°C; one tube at 4°C for use.
   - **(ii)** PBS with 2 × penicillin-streptomycin: 10 mL 100 × penicillin-streptomycin; add into 395 mL DMEM/F12, mix, and store in 4°C.
   - **(iii)** Culture medium: 100 mL:15 mL FBS, 1 mL 200 mM GlutaMAX, and 1 mL 100 × penicillin-streptomycin; add into 83 mL DMEM, mix, and store at 4°C.

(2) Add 10 mL adipose tissue from liposuction into 50 mL tube.

(3) Wash adipose tissue three times by PBS.

(4) Prepare three 100 mm cell culture dishes with 5–10 mL PBS (2 × penicillin-streptomycin) in them, disinfect the tissue by dip it in the dish successively, immerse 3 min in every dish.

(5) Put the tissue in the 10 cm dish with 5–10 mL culture medium in it, remove other kinds of tissue (such as connective tissue and blood vessel), and cut it up into 1 cm × 1 cm × 1 cm pieces by surgical tools, and then cut it up into pieces as small as possible (smaller than 1 mm^3 every pieces).

(6) Move minced tissue into 50 mL tube with collagenase solution (tissue of white zone using collagenase type I (0.5%) solution and trypsin-EDTA (0.5%), the tissue of red zone using collagenase solution of type I). The amount of collagenase solution should be more than 10-fold of the volume of tissue. Put the tube into water bath shaker, 37°C, 80 rpm for 60–120 min. During digesting, shake the tube lightly every 5–15 min (standard: digestion until the tissue becomes thick liquid; no obviously spot adjunct).

(7) Add FBS with the amount of 5% of the total volume to neutralize the reaction and mix the solution for 15–20 times.

(8) Put 70 μm cell strainer on a new 50 mL tube and add the digested solution on it slowly to the filtrate. Wash the tube of digestant solution tube with culture medium, the liquid filtrated by cell strainer (filtration residue can be collected and cultured by culture medium overnight).

(9) Re-suspend the precipitate by 10 mL culture medium, centrifuge by 400 × g/5 min, and discard the supernatant.

(10) Re-suspend the precipitate by 10 mL culture medium (DMEM/F12 medium with 1 × penicillin-streptomycin/10% FBS) and count the cell.

(11) Plank the cell in the flask with the density of 0.5–1 × 10^6 cells/T25. If the amount of cells is too little, cells can be cultured by the cell culture plate. Culture at 37°C, 5%CO_2.

2.3.2 Cell culture and freezing

Reagent:

Culture medium (1 × penicillin-streptomycin/20% FBS DMEM/F12).
Freezing medium: FBS:DMSO = 9:1.
Trysin.
PBS.

Materials:

T25 flask, serological pipettes (5, 10, 25 mL), 1.5 mL centrifuge tubes, 15/50 mL tubes, pipette tips (10, 200, 1000 μL), freezing tubes, liquid nitrogen tank, cell cryopreservation box.

Equipment:

Liquid nitrogen container, Refrigerator (4°C and −20°C).

Procedure:

(1) Culture and passage.
  (i) Add half amount of culture medium after primary cells have been planked 24–48 h.
  (ii) After adding culture medium at 36–48 h, change medium and observe cells by microscope. When the cell confluence reaches 80%, detach cells by trypsin and then digest and passage cells. Plank the cell in the flask with the density of 0.25–0.5 × 10^6 cells/T25.
(2) Cell freezing:
  (i) Freeze the cells by the freezing medium of FBS: DMSO = 9:1.
  (ii) Detach cells by trypsin and neutralize by culture medium with the same amount. Centrifuge by 400 × *g*/5 min, re-suspend cell by 5 mL culture medium, and count the cell. Re-suspend the cells by freezing medium at the cell concentration of 5 × 10^5 cells/mL; every freezing tube contains 1–2 mL.

2.4 Corneal stem cells

2.4.1 Isolation of primary corneal stem cells

Reagent:

DMEM/F12 medium, PBS, FBS, Trypsin-EDTA (0.5%), dispase, 100×penicillin-streptomycin, 100 × GlutaMAX (200 mM).

Materials:

70 μm cell strainers, 50 mm cell culture dishes, T25 flasks, serological pipettes (5, 10, 25 mL), 1.5 mL centrifuge tubes, 15/50 mL centrifuge tubes, pipette tips (10, 200, 1000 μL).

Surgical tools:

Ophthalmic forceps, ophthalmic scissors, forceps (20–35 cm), tissue scissors (straight head, cured head), scalpels.

Equipment:

Operation microscope; centrifuge; biosafety cabinet; water bath; refrigerator (4°C and −20°C).

Tissue source:

Aborted fetal.

Procedure:

(1) Preparation for solution and reagent:
   (i) FBS: Make it melt at 4°C, divide into 50 mL tube, 45 mL every tube, store at −20°C, one tube at 4°C for use.
   (ii) PBS with 2 × penicillin-streptomycin: 10 mL 100 × penicillin-streptomycin; add into 395 mL DMEM/F12, mix, and store at 4°C.
   (iii) Culture medium: 100 mL:15 mL FBS, 1 mL 100 × GlutaMAX, and 1 mL 100 × enicillin-streptomycin; add into 83 mL DMEM, mixing and store at 4°C.
   (iv) Dispase dilution (2 U/mL): dissolve dispase by DMEM/F12 medium at the concentration of 2 U/mL.
(2) Separate limbal tissue from aborted fetal by surgical tools.
(3) 50 mm dishes with 5–10 mL PBS (2 × penicillin-streptomycin).
(4) Wash limbal tissue three times by PBS.
(5) Move the tissue into the cell culture dish with dispase dilution (2 U/mL) in it; digest for 1 h.
(6) Remove the epithelium, move another part of tissue to 25 cm flask with culture medium in it, culture under 37°C, 5%CO_2.
(7) Change the culture medium every 48 h, until the cell confluence reaches 80% passage the cells.

2.4.2 Cell culture and freezing

Reagent:

Culture medium (1 × penicillin-streptomycin/20% FBS DMEM/F12).
Freezing medium: FBS:DMSO = 9:1.
Trypsin.
PBS.

Materials:

T25 flask, serological pipettes (5, 10, 25 mL), 1.5 mL EP tubes, 15/50 mL centrifuge tubes, pipette tips (10, 200, 1000 μL), freezing tubes, liquid nitrogen, cell cryopreservation box.

Equipment:

Liquid nitrogen container, Refrigerator (4°C and −20°C).

Procedure:

(1) Culture and passage.
   (i) Add half amount of culture medium after primary cells have been planked at 24–48 h.

(ii) After adding culture medium for 36–48 h, change medium and observe cells by microscope. When the cell confluence reaches 80%, detach cells by Trypsin and then digest and passage cells. Plank the cell in the flask with the density of 0.25–0.5 × 10^6 cells/T25.

(2) Cell freezing:
   (i) Freeze the cells by the freezing medium of FBS:DMSO = 9:1.
   (ii) Detach cells by Trypsin and neutralize by culture medium with the same amount. Centrifuge by 400 × *g*/5 min, re-suspend cell by 5 mL culture medium, and count the cell. Re-suspend the cells by freezing medium at the cell concentration of 5 × 10^5 cells/mL; every freezing tube contains 1–2 mL.

2.5 Identification of MSCs

2.5.1 Flow cytometry identification

Reagent, material, equipment:

Flow cytometer, Refrigerator (4°C and −20°C).

Mouse antihuman fluorescein isothiocyanate-labeled/PE-labeled monoclonal:

For BMSC, hUCB-MSCs, and ADSC: CD14, CD34, CD45, CD73, CD90, CD105.

For corneal stem cells: ABCG2, ΔNp63, cytokeratin, integrin.

Procedure:

(1) Seed the cells at a 6-well plate with the density of 2 × 10^6 cells/well.
(2) Detach P3 cells by trypsin-EDTA (0.25%) cells with the density 80%–90%; neutralized by culture medium with the same amount.
(3) Wash cells twice by PBS.
(4) Adjust the cell concentration to 1 × 10^7 cells/mL and transfer 100 μL of the suspension into a 1.5 mL tube.
(5) Block the cells by 50 μL normal goat serum (Gibco) for 15–20 min at room temperature.
(6) Incubate the cells with monoclonal anti-CD14, anti-CD34, anti-CD45, anti-CD73, anti-CD90, anti-CD105 (dilution, 1:50), and kept at 4°C for 20 min.
(7) Wash the cells twice by PBS to remove the excess antibody and add 500 μL fixative solution.
(8) Analyze cells using a BD FACS Calibur (BD Biosciences) and perform quantification by Cell Quest software (version 3.3; BD Biosciences). Use FITC- or PE-labeled immunoglobulin G-stained cells as negative controls.

Result:

For BMSC and UMSC:

Positive: CD73, CD90, CD105.

Negative: CD14, CD34, CD45.

For corneal stem cells:

Positive: ABCG2, *ΔNp63*, cytokeratin, integrin.

2.5.2 RNA extraction and qPCR

Reagent, material, equipment:

TRIzol reagent
Primers
Rnase-free water
Chloroform
SuperScript III Reverse Transcriptase
SYBR green master mix
MircroAmp Optical 96-well reaction plate (RNase-free, Biosystems)
Strip tube (RNase-free, Axygen)
0.5 mL tubes
1.5 mL tubes
200 μL tips (Rnase-free, Axygen)
10 μL tips (Rnase-free, Axygen)
Microcentrifuge tube racks
Icebox (hole of 0.2, 0.5, 1.5, 2 mL)
Real-time PCR cycler

Procedure:

(1) Culture the cells until 1×10^7.
(2) Add 1 mL TRIZOL Reagent.
(3) Transfer the solution to a 1.5 mL tube and allow it to rest at RT (room temperature) for 15–30 min to lyse the cells.
(4) For 15 min at 2°C to 8°C, centrifuge at $12{,}000 \times g$ after adding chloroform (0.2 mL) and vortexing for 15 s.
(5) Separate the mixture following centrifugation into an interphase, a colorless upper aqueous phase, and a lower red phenol-chloroform phase. The RNA will be completely stable in the liquid state. Without disturbing the interphase, the aqueous phase (upper) should be transferred to a new tube. Calculate the aqueous phase's volume (approximately 60% of the TRIZOL Reagent's volume during homogenization).
(6) Mix isopropyl alcohol with the aqueous phase to precipitate the RNA. For the initial homogenization, apply isopropyl alcohol (0.5 mL) to TRIZOL Reagent (1 mL). Centrifuge samples for 10 min at 2°C to 4°C at a maximum of $12{,}000 \times g$, following the sample's incubation for 10 min at 15°C to 30°C. The RNA residue, typically indistinguishable, produces a gel-like pellet on the side and bottom of the tube before centrifugation.
(7) Afterwards, eliminate the supernatant. At least 75% ethanol (~1 mL) should be poured for the TRIZOL Reagent (every 1 mL) employed for the first homogenization to wash the RNA pellet thoroughly. Vortex the samples together, then centrifuge at 2°C to 8°C for 5 min at a maximum of $7500 \times g$. Continue the previously described rinsing method, and eliminate any remaining ethanol.
(8) Dry the RNA pellet in the vacuum or air for 5–10 min. Do not dry the RNA pellet before centrifuging it in a vacuum. The RNA pellet must not be allowed

to dry since this will substantially decrease its solubility ultimately. During passing the solution through a pipette tip a few times, the RNA should be dissolved in DEPC-treated water. The A260/A280 ratio of partially dissolved RNA samples is <1.6 usually.

(9) Finally, RNA (1 μL) should be diluted with DEPC-treated water (39 μL and 1:40 dilution). Through a microcuvette (10 μL), determine the optical density (OD) at 260–280 nm to calculate the concentration and purity of the sample. Applying the standard that 40 g/mL RNA equals 1 OD at 260, the A260/A280 ratio should be >1.6.

(10) qPCR procedure:

- **(i)** Set the PCR procedure.
- **(ii)** Design the 96-well reaction plate.
- **(iii)** Calculate the amount of all the reagent base on the amount.
- **(iv)** Pipette Probe qPCR Mix, primers, probes, and DD Water into 0.5 mL tube.
- **(v)** Divide the mix into a 96-well reaction plate.
- **(vi)** Cover the obsolete pipette DNA template to wells.
- **(vii)** Move the plate to the qPCR cycler, and begin the reaction.
- **(viii)** Record the result.

The reaction mixture (example)

Reagent	Reaction volume
Rnase-free/Dnase-free water	6.5 μL
Reverse primer	1 μL
Forward primer	1 μL
Template DNA	5 μL
SYBR Green qPCR Mix	12.5 μL
Total volume	25 μL

PCR procedure (example)

94°C	8 min	
94°C	3 s	} 35 ×
52°C	15 s	
45°C	90s	

3. Hematopoietic stem cells

3.1 Apheresis

3.1.1 Isolation of primary umbilical cord blood MSCs

Reagent

Percoll (1.073 g/mL), 4% sodium citrate, HSC culture medium, PBS, FBS, trypsin, trypsin-EDTA (0.25%), penicillin-streptomycin liquid (100×).

Materials

T25 flasks, pipette (25, 10, and 5 mL), 1.5 mL EP tube, 15/50 mL tube, pipet tips (1000, 200, and 10 μL), 50 mL tubes.

Equipment

Centrifuge; biosafety cabinet; water bath; refrigerator (4°C and −20°C).

Procedure

(1) Inject hematopoietic stem cell mobilizing agents G-CSF to the donor, draw bone marrow.
(2) First, pour 2 mL of 4% sodium citrate in a tube (50 mL).
(3) Add bone marrow (10 mL) to a 50 mL tube, and invert to mix gently.
(4) Dilute the bone marrow with PBS (v/v; 1:1): add PBS (12 mL) to a tube (50 mL).
(5) Prepare 50 mL tubes with 15 mL of Percoll, and slowly pour diluted bone marrow on top of the Percoll. Blood must be added slowly and stay on top of the Percoll.
(6) Centrifuge with the BRAKE OFF and Accelerator at RT at $400 \times g$ for 20 min.
(7) Collect the MNC layer through a new plastic transfer pipette and a 50 mL vial after centrifugation.
(8) MNC suspension (obtained in step-7) should then be diluted into 15–20 mL of each tube in a falcon tube (50 mL), diluted 1:1 in PBS, and centrifuged at RT for 20 min at $300 \times g$ with the accelerator and BRAKE ON.
(9) By introducing 1 mL of enriched HSC growth media to each pellet and pipetting up and down constantly, eliminate the supernatant and resuspend the cell pellets.
(10) Fill the tube with 9 mL of the HSC culture medium and frequently pipet it up and down.
(11) Count the cells, and seed the cells by the density of $2 \times 10^5/cm^2$. Culture under 37°C, and 5% CO_2.

3.1.2 Cell culture and freezing

Reagent

HSC culture medium
Freezing medium: FBS:DMSO = 9:1
Trypsin
PBS

Materials

T25 flasks, 25, 10, and 5 μL of pipette, EP tube (1.5 mL), 15/50 mL tube, pipet tips (10, 200, 1000 μL), 50 mL tubes, freezing tube, liquid nitrogen, cell cryopreservation box.

Equipment

Liquid nitrogen container
Refrigerator (4°C and −20°C)

Procedure

(1) Culture and passage.

  (i) Add half the amount of culture medium after primary cells have been planked 24–48 h.

  (ii) After adding culture medium 36–48 h, change medium and observe cells by microscope. When the cell confluence reaches 80%, detach cells by trypsin and then digest and passage cells. Plank the cell in the flask by the density of $0.25–0.5 \times 10^6$/T25.

(2) Cell freezing

  (i) Freeze the cells with the freezing medium of FBS:DMSO = 9:1

  (ii) Detach cells with trypsin and neutralize by culture medium with the same amount. Resuspend the cell in a culture medium (5 mL), then for 5 min, perform the centrifugation at $400 \times g$, and count the cells. Freeze medium at a cell concentration of 5×10^5/mL to resuspend the cells. Every freezing tube contains 1–2 mL.

3.1.3 Flow cytometry identification

Reagent, material, equipment

Flow cytometer
Refrigerator (4°C and −20°C)
Refrigerated centrifuge
Mouse antihuman PE-labeled/FITC-labeled monoclonal anti-bodies: CD34, CD59, CD90, CD117, CD38, EMCN, and lineage cocktail 1 (lin-1)

Procedure

(1) 2×10^6 cells/well are used to plant the cells in a 6-well plate.

(2) Detach P3 cells by trypsin-EDTA (0.25%) cells at 80%–90% density, neutralized by culture medium with the same amount.

(3) Wash cells twice by PBS.

(4) Transfer 100 μL of the mixture into a tube (1.5 mL) after adjusting the cell density to 1×10^7 cells/mL.

(5) After that, block the cells for 15–20 min at RT with 50 μL of normal goat serum (Gibco).

(6) Kept the cells in a 20-min incubation period with the monoclonal antibody at 4°C.

(7) After two PBS washes to remove any remaining antibody, add 500 μL of the fixative solution to the cells.

(8) Subsequently, employ a BD FACS Calibur for cell analysis and Cell Quest software v.3.3 (BD Biosciences) for quantification. As negative controls, use immunoglobulin G-stained cells PE-/FITC-labeled.

Results

Positive: CD34, CD59, CD90, CD117, CD38, EMCN.
Negative: CD38, lin.

4. Materials, reagents, chemicals, apparatus, and kits

4.1 Thermoresponsive poly(butylacrylate-*co*-*N*-isopropylacrylamide)-coated interface

The materials including reagents, chemicals, and kits for preparation of thermoresponsive poly(butylacrylate-*co*-*N*-isopropylacrylamide)-coated (PNB) interface are shown in the following Table 1.

4.2 Procedure

4.2.1 Preparation process of PNB-coated interface

The preparation process of thermoresponsive poly(butylacrylate-*co*-*N*-isopropylacrylamide)-coated (PNB) interface is described as one example of preparation method of thermoresponsive interface as follows.

Poly(butylacrylate-*co*-*N*-isopropylacrylamide), PNB (12% of polybutylacrylate and 88% of poly-*N*-isopropylacrylamide; molecular weight (Mw) = 30,000), is dissolved in 99.5% ethanol to generate concentration of 8 mg/mL to be utilized as the PNB coating solution. A total of 600 μL of PNB coating solution is added to each well of 12-well tissue culture polystyrene (TC) plates (surface area = 4 cm^2/well) and then is incubated for 120 min at 37°C. Then, the plates are rinsed with ethanol four times and subsequently rinsed with pure water four times; PNB-TC plate indicates a PNB-coated TC plate. For hESC and hiPSC cultivation, PNB-TC plates or TC plates are further coated with 5 μg/mL recombinant vitronectin (rVN) for 120 min at 37°C. Subsequently, the rVN solution is discarded, and rVN-coated PNB-TC plates (rVN-PNB-TC) or rVN-coated TC plates (rVN-TC) are utilized for hESC and hiPSC cultivation.

4.2.2 Characterization of PNB-coated interface

X-ray photoelectron spectroscopy (XPS) is utilized for the analysis of the chemical composition of the surface on the thermoresponsive PNB-TC and rVN-PNB-TC plates. The binding energy (BE) scale is standardized by positioning 284.6 eV for the peak maximum of the C1s spectra.

The water contact angle of the thermoresponsive PNB-TC and rVN-PNB-TC plates is measured for the evaluation of hydrophilicity on the surface of thermoresponsive PNB-TC and rVN-PNB-TC plates utilizing the sessile drop process with a contact angle goniometer at 37°C on a hot copper plate as well as 7–9°C on a cooled copper plate after 0.5 h.

4.2.3 Preparation and culture of hESCs and hiPSCs

Human iPSCs and hESCs are cultivated on rVN-coated plates in Essential 8 (E8) media as conventional method (Peng et al., 2015; Sunga et al., 2019). hESCs or hiPSCs are inoculated on rVN-PNB-TC plates in E8 media at a seeding density of 5×10^4 cells/cm^2, and the media is replaced daily with fresh media during hESCs and hiPSCs cultivation. After hESCs and hiPSCs culture for 5–7 days to approximately 78%–88% confluence, hESCs and hiPSCs are partially released from the

Table 1 Materials used in this study.

Materials	Abbreviation	Catalog No.	Company
ECM			
Matrigel	Matrigel	#356230	CO
Recombinant vitronectin	rVN	A14700	TFS
Cell culture dishes			
Six-well polystyrene plate	TCP	#353046	CO
Chemicals and polymer			
Dispase II	Dispase	D4693-1G	SA
Poly(butylacrylate-*co*-*N*-isopropylacrylamide)	PNB	762881-5G	SA
Cell culture medium and component			
Essential 8 medium	Essential 8	A1517001	TFS
Essential 6 medium	Essential 6	A1516401	TFS
DMEM/F12 medium	DMEM/F12	11330-057	TFS
RPMI 1640	RPMI 1640	11875093	TFS
Alkaline phosphatase assay kit	SensoLyte pNPP	AS-72146	AS
Hoechst 33342	Hoechst	PA-3014	LO
Antibodies			
Anti-Oct3/4 antibody	Anti-Oct3/4 antibody	sc-5279	SCB
Anti-Sox2 antibody	Anti-Sox2 antibody	AB5603	ME
Anti-SSEA-4 antibody	Anti-SSEA-4 antibody	ab16287	AB
Anti-Nanog antibody	Anti-Nanog antibody	MA1-017	TFS
Mouse IgG1 Isotype antibody	Isotype-control	MA5-14453	TFS
Alexa Fluor 488 goat antimouse IgG	Alexa Fluor 488 goat antimouse IgG	A-11001	TFS
Alexa Fluor 488 goat antirabbit IgG	Alexa Fluor 488 goat antirabbit IgG	A-11008	TFS
Alexa Fluor 594 goat antimouse IgG	Alexa Fluor 555 goat antimouse IgG	A-21424	TFS
Mice			
NOD.CB17-Prkdcscid/Jnarl	NOD-SCID mice	NOD.CB17-Prkdcscid/Jnarl	NLAC

CO, Corning; TFS, Thermo Fisher Scientific; SA, Sigma-Aldrich; AS, AnaSpec; LO, Lonza; SCB, Santa Cruz Biotechnology; ME, Merck KGaA; AB, Abcam; NLAC, National Laboratory Animal Center.

thermoresponsive surface of the rVN-PNB-TCP plates using the following processes. First, the thermoresponsive rVN-PNB-TCP plates are placed on a copper plate (0.5 cm (height) × 11.5 cm (length) × 7.5 cm (width)) at 4°C for 30 min. Then, the cultivation media on the thermoresponsive rVN-PNB-TCP plates are gently pipetted twice with a 1-mL pipette to effectively release hESCs and hiPSCs. After the removal of the partially released hESCs and hiPSCs from the rVN-PNB-TCP plates, fresh E8 media are added to the rVN-PNB-TCP plates, and the residual cells on the plates are repeatedly cultivated in a CO_2 incubator at 37°C. This process can be performed for five cycles on thermoresponsive rVN-PNB-TCP plates. The cultivation parameters of hESCs and hiPSCs such as the detachment rate, differentiation rate, and expansion fold change are evaluated.

During each cycle of hESCs and hiPSCs cultivation, the following equation is used to characterize the expansion fold change (Peng et al., 2015; Sunga et al., 2019).

$$\text{Expansion fold change} = \text{The number of hESCs and hiPSCs on the plates after cultivation} / (\text{The number of hESCs and hiPSCs on the plates before cultivation}) \quad (1)$$

The following equation describes the evaluation of the differentiation rate for the hESCs and hiPSCs colonies after cultivation on the thermoresponsive rVN-PNB-TCP plates before detachment of the hESCs and hiPSCs colonies (Peng et al., 2015; Sunga et al., 2019):

$$\text{Differentiation rate } (\%) = \text{Percentage of differentiated colonies } (\%) + \text{Percentage of partially differentiated colonies} \times 0.5\ (\%) \quad (2)$$

where partially differentiated colonies are denoted as colonies with alkali phosphatase (AP) activities in the center of the colonies but not at the edge of the colonies, and differentiated colonies are denoted as colonies without AP activity.

The following equation characterizes the detachment rate of hESCs and hiPSCs from the thermoresponsive rVN-PNB-TCP plates (Peng et al., 2015; Sunga et al., 2019):

$$\text{Detachment rate}\% = (\text{The number of hESCs and hiPSCs released from the plates}) / \text{The number of hESCs and hiPSCs on the plates before release} \times 100\% \quad (3)$$

5. Conclusion

The isolation, culture, and freezing of adult stem cells have always been one of the key points in the field of cell therapy and regenerative medicine. Still, due to the difficulty of establishing uniform standards for experimentation and production, different research results and product quality have been obtained, which is one of the main obstacles to the application of human adult stem cells in regenerative medicine. Although the research on the experimental techniques of human adult stem cells is not yet complete, researchers are constantly working to improve the techniques. We believe that in the future there will be better ways to obtain homogeneous and highly active human adult stem cells, whose biological properties such as self-replication, multidirectional differentiation, immune resistance, and secretory effects will play a more significant role in the treatment of clinical diseases. This chapter provides a protocol for the isolation, culture, and freezing techniques of human adult stem cells, hopefully offering researchers in the field a macroscopic understanding and a reference for specific operational details.

Abbreviations

3D	three-dimensional
ABCG2	ATP-binding cassette super-family G member 2
ADSCs	adipose tissue-derived stromal cells
AP	alkali phosphatase
BMSCs	bone marrow mesenchymal stem cells
cRGDfK	cyclic Arg-Gly-Asp-Phe-Lys
DMSO	dimethyl sulfoxide
EB	embryoid body
FBS	fetal bovine serum
GRGDS	Gly-Arg-Gly-Asp-Ser
hESCs	human embryonic stem cells
hiPSCs	human induced pluripotent stem cells
HLA-DR	human leukocyte antigen-DR isotype
HSCs	hematopoietic stem cells
hUCB-MSCs	human umbilical-cord-blood-derived mesenchymal stem cells
IL-11	interleukin 11
IL-6	interleukin 6
LCST	lower critical solution temperature
LIF	leukemia inhibitory factor
MC	methylcellulose
M-CSF	macrophage colony-stimulating factor
MMA	methyl methacrylate
MNC	mononuclear Cells
MS	mesenchymal stem
MSCs	mesenchymal stem cells
NSP	nitrobenzospiropyran
PBS	phosphate buffered saline
PEG	poly(ethylene glycol)
PEO	polyethylene oxide
PMMA	polymethylmethacrylate
PNB	poly(butylacrylate-co-N-isopropylacrylamide)

PNIPA	poly(N-isopropyl acrylamide)
PPO	polypropylene oxide
RGD	Arg-Gly-Asp
RGDS	Arg-Gly-Asp-Ser
rVN	recombinant vitronectin
SAM	self-assembled monolayer
TC	tissue culture polystyrene
UV	ultraviolet
VIS	visible light
XPS	X-ray photoelectron spectroscopy

References

Auernheimer, J., Dahmen, C., Hersel, U., Bausch, A., Kessler, H., 2005. Photoswitched cell adhesion on surfaces with RGD peptides. J. Am. Chem. Soc. 127, 16107–16110.

Brun, A., Richard, C., Bessodes, M., Scherman, D., Merten, O.-W., 2010. Thermoresponsive surfaces for cell culture and enzyme-free cell detachment. Prog. Polym. Sci. 35, 1311–1324.

Delaittre, G., Greiner, A., Paulöhrl, T., Bastmeyer, M., Barner-Kowollik, C., 2012. Correction: chemical approaches to synthetic polymer surface biofunctionalization for targeted cell adhesion using small binding motifs. Soft Matter 8, 7323–7347.

Dulak, J., Szade, K., Szade, A., Nowak, W., Józkowicz, A., 2015. Adult stem cells : hopes and hypes of regenerative medicine *. Acta Biochim. Pol. 62, 329–337.

Ebara, M., Yamato, M., Aoyagi, T., Kikuchi, A., Sakai, K., Okano, T., 2004a. Temperature-responsive cell culture surfaces enable "on – off" affinity control between cell integrins and RGDS ligands. Biomacromolecules 5, 505–510.

Ebara, M., Yamato, M., Aoyagi, T., Kikuchi, A., Sakai, K., Okano, T., 2004b. Immobilization of cell-adhesive peptides to temperature-responsive surfaces facilitates both serum-free cell adhesion and noninvasive cell harvest. Tissue Eng. 10, 1125–1135.

Economescu, M., Rubach, M., 2015. Hematopoietic stem cells therapies. Curr. Stem Cell Res. Ther. 12.

Ekerdt, B., Fuentes, C., Lei, Y., Adil, M., Ramasubramanian, A., Segalman, R., Schaffer, D., 2018. Thermoreversible hyaluronic acid-PNIPAAm hydrogel systems for 3D stem cell culture. Adv. Healthc. Mater. 7, 1800225.

George, J., Manjusha, W.A., Jegan, S.R., Mahija, S.P., Josphin, J.S., 2018. A review of stem cells in regenerative medicine. Int. J. Sci. Res. Sci. Technol. 3, 806–815.

Gong, Y.-H., Yang, J., Cao, F.-Y., Zhang, J., Cheng, H., Zhuo, R.-X., 2013. Photoresponsive smart template for reversible cell micropatterning. J. Mater. Chem. B 1, 2013–2017.

Higuchi, A., Hamamura, A., Shindo, Y., Kitamura, H., Yoon, B., Mori, T., Uyama, T., Umezawa, A., 2004. Photon-modulated changes of cell attachments on poly(spiropyran- co - methyl methacrylate) membranes. Biomacromolecules 5, 1770–1774.

Higuchi, A., Yamamoto, T., Sugiyama, K., Hayashi, S., Tak, T.M., Nakagawa, T., 2005. Temperature-dependent cell detachment on pluronic gels. Biomacromolecules 6, 691–696.

Higuchi, A., Aoki, N., Yamamoto, T., Miyazaki, T., Fukushima, H., Tak, T.M., Jyujyoji, S., Egashira, S., Matsuoka, Y., Natori, S., 2006. Temperature-induced cell detachment on immobilized Pluronic surface. J. Biomed. Mater. Res. A 79, 380–392.

Higuchi, A., Ling, Q.-D., Kumar, S., Chang, Y., Kao, T.-C., Murugan, A., Alarfajj, A., Hsu, S.-T., Umezawa, A., 2014. External stimulus-responsive biomaterials designed for the culture and differentiation of ES, IPS, and adult stem cells. Prog. Polym. Sci. 39.

Huang, C.-C., Liao, C.-K., Yang, M.-J., Chen, C.H., Hwang, S.-M., Hung, Y.-W., Chang, Y., Sung, H.-W., 2010. A strategy for fabrication of a three-dimensional tissue construct containing uniformly distributed embryoid body-derived cells as a cardiac patch. Biomaterials 31, 6218–6227.

Kaneko, S., Nakayama, H., Yoshino, Y., Fushimi, D., Yamaguchi, K., Horiike, Y., Nakanishi, J., 2011. Photocontrol of cell adhesion on amino-bearing surfaces by reversible conjugation of poly(ethylene glycol) via a photocleavable linker. Phys. Chem. Chem. Phys. 13, 4051–4059.

Kawamura, M., Miyagawa, S., Miki, K., Saito, A., Fukushima, S., Higuchi, T., Kawamura, T., Kuratani, T., Daimon, T., Shimizu, T., Okano, T., Sawa, Y., 2012. Feasibility, safety, and therapeutic efficacy of human induced pluripotent stem cell-derived cardiomyocyte sheets in a porcine ischemic cardiomyopathy model. Circulation 126, S29–S37.

Kim, D., Choi, A., Seo, K., Yoon, H., Han, S., 2019. Bulk poly(N-isopropylacrylamide) (PNIPAAm) thermoresponsive cell culture platform: toward a new horizon in cell sheet engineering. Biomater. Sci 7.

Klump, H., 2020. Hematopoietic. Stem Cells, 1–19.

Lachaud, C., Hmadcha, A., Soria, B., 2019. Corneal stem cells: identification and methods of ex vivo expansion: therapy and surgery. In: L. Alió, J., et al. (Eds.), Corneal Regeneration. Springer, pp. 57–75.

Lee, C.S., Burnsed, O.A., Raghuram, V., Kalisvaart, J., Boyan, B.D., Schwartz, Z., 2012. Adipose stem cells can secrete angiogenic factors that inhibit hyaline cartilage regeneration. Stem Cell Res Ther 3, 35.

Li, Q.-Q., Wu, Z.-Q., Zhang, L.-J., 2017. Directional differentiation of bone marrow mesenchymal stem cells. Chinese J. Tissue Eng. Res. 21, 4082–4087.

Liu, D., Xie, Y., Shao, H., Jiang, X., 2009. Using azobenzene-embedded self-assembled monolayers to photochemically control cell adhesion reversibly. Angew. Chem. Int. Ed. Engl. 48, 4406–4408.

Masumoto, H., Matsuo, T., Yamamizu, K., Uosaki, H., Narazaki, G., Katayama, S., Marui, A., Shimizu, T., Ikeda, T., Okano, T., Sakata, R., Yamashita, J., 2012. Pluripotent stem cell-engineered cell sheets reassembled with defined cardiovascular populations ameliorate reduction in infarct heart function through cardiomyocyte-mediated neovascularization. Stem Cells 30, 1196–1205.

Matsuura, K., Okano, T., 2013. Cell sheet transplantation for heart tissue repair. J. Control. Release, 169.

Matsuura, K., Masuda, S., Haraguchi, Y., Yasuda, N., Shimizu, T., Hagiwara, N., Zandstra, P., Okano, T., 2011. Creation of mouse embryonic stem cell-derived cardiac cell sheets. Biomaterials 32, 7355–7362.

Matsuura, K., Wada, M., Konishi, K., Sato, M., Iwamoto, U., Sato, Y., Tachibana, A., Kikuchi, T., Iwamiya, T., Shimizu, T., Yamashita, J., Yamato, M., Hagiwara, N., Okano, T., 2012a. Fabrication of mouse embryonic stem cell-derived layered cardiac cell sheets using a bioreactor culture system. PLoS One 7, e52176.

Matsuura, K., Wada, M., Shimizu, T., Haraguchi, Y., Sato, F., Sugiyama, K., Konishi, K., Shiba, Y., Ichikawa, H., Tachibana, A., Ikeda, U., Yamato, M., Hagiwara, N., Okano, T., 2012b. Creation of human cardiac cell sheets using pluripotent stem cells. Biochem. Biophys. Res. Commun. 425, 321–327.

Nandivada, H., Ross, A., Lahann, J., 2010. Stimuli-responsive monolayers for biotechnology. Prog. Polym. Sci. 35, 141–154.

Park, I.-S., Chung, P.-S., Ahn, J.C., 2015. Adipose-derived stromal cell cluster with light therapy enhance angiogenesis and skin wound healing in mice. Biochem. Biophys. Res. Commun. 462, 171–177.

Peng, I.-C., Chih-Chen, Y., Lu, Y.-T., Muduli, S., Ling, Q.-D., Alarfaj, A., Murugan, A., Kumar, S., Murugan, K., Lee, H., Chang, Y., Higuchi, A., 2015. Continuous harvest of stem cells via partial detachment from thermoresponsive nanobrush surfaces. Biomaterials 76, 76–86.

Ross, K., Veltri, L., Kanate, A., 2019. Umbilical cord blood transplantation. In: Bashir, Q. (Ed.), Hematopoietic Cell Transplantation for Malignant Conditions. Elsevier, pp. 125–137.

Sakaguchi, K., Shimizu, T., Horaguchi, S., Sekine, H., Yamato, M., Umezu, M., Okano, T., 2013. In vitro engineering of vascularized tissue surrogates. Sci. Rep. 3, 1316.

Sasamoto, Y., Oie, Y., Nishida, K., 2019. Corneal stem cell-based therapies: therapy and surgery. In: Corneal Regeneration. Springer, pp. 155–172.

Schiesser, W., 2019. Hematopoietic. Stem Cells, 179–194.

Shi, S., Gronthos, S., 2013. Bone marrow mesenchymal stem cells. In: Stem Cells in Craniofacial Development and Regeneration, pp. 223–239.

Stabler, C., Lecht, S., Lazarovici, P., Lelkes, P., 2015. Mesenchymal stem cells for therapeutic applications in pulmonary medicine. Br. Med. Bull 115.

Sunga, T.-C., Yang, J.-S., Chih-Chen, Y., Liu, Y.-C., Jiang, Y.-P., Lu, M.-W., Ling, Q.-D., Kumar, S., Chang, Y., Umezawa, A., Chen, H., Higuchi, A., 2019. The design of a thermoresponsive surface for the continuous culture of human pluripotent stem cells. Biomaterials 221, 119411.

Tm, L., 2019. Stemness of mesenchymal stem cells. J. Stem Cell Ther. Transplant., 1–4.

Weihua, H., Cheng, S.-X., Zhuo, R.-X., 2009. Thermo-sensitive polymeric micelles based on poly(N-isopropylacrylamide) as drug carriers. Prog. Polym. Sci. 34, 893–910.

Wu, L.-H., Chen, T., Wang, R., 2016. Bone marrow mesenchymal stem cell transplantation for osteoarthritis. Chinese J. Tissue Eng. Res. 20, 6782–6787.

Yang, L., Fan, X., Zhang, J., Ju, J., 2020. Preparation and characterization of thermoresponsive poly(N-isopropylacrylamide) for cell culture applications. Polymers (Basel) 12, 389.

Zhang, R., Mjoseng, H.K., Hoeve, M.A., Bauer, N.G., Pells, S., Besseling, R., Velugotla, S., Tourniaire, G., Kishen, R.E.B., Tsenkina, Y., Armit, C., Duffy, C.R.E., Helfen, M., Edenhofer, F., de Sousa, P.A., Bradley, M., 2013. A thermoresponsive and chemically defined hydrogel for long-term culture of human embryonic stem cells. Nat. Commun. 4, 1335.

CHAPTER

4 Isolation of human stem cells from full-term gestation amniotic fluid using immunoselection and one-stage culture techniques

Norshariza Nordin[a,b], Khairul Akmal Abdul Rahman[a], Nurfarhana Ferdaos[c], Adila A. Hamid[d], Muhammad Khair Joharry[a], Zulida Rejali[e], Thilakavathy Karuppiah[a,b], and Mohd Nazri Yazid[e]

[a]*Department of Biomedical Science, Faculty of Medicine and Health Sciences, Universiti Putra Malaysia, Serdang, Selangor, Malaysia,* [b]*Genetics & Regenerative Medicine (ReGEN) Research Group, Faculty of Medicine and Health Sciences, Universiti Putra Malaysia, Serdang, Selangor, Malaysia,* [c]*Department of Pharmacology and Chemistry, Faculty of Pharmacy, Universiti Teknologi MARA Puncak Alam Campus, Selangor, Malaysia,* [d]*Department of Physiology, Faculty of Medicine, National University of Malaysia Medical Centre, Kuala Lumpur, Malaysia,* [e]*Department of Obstetrics and Gynaecology, Faculty of Medicine and Health Sciences, Universiti Putra Malaysia, Serdang, Selangor, Malaysia*

1. Introduction

Stem cells are promising cells for various therapeutic applications, including cell replacement therapy. Their ability to undergo self-renewability permits them to be preserved as stem cell lines for prolonged culture. Nowadays, researchers have identified the presence of potent stem cells residing in the human amniotic fluid, namely as amniotic fluid-derived stem cells (Zentelytė et al., 2021). It has been shown to exhibit intermediate properties between embryonic and adult stem cells in that they have broader differentiation potential but with no ability to form teratoma in vivo. Recently, an increasing number of research groups have been interested in isolating stem cells from the amniotic fluid of full-term gestation. This is due to their procurement during delivery, which devoids them from being involved in an invasive technique, amniocentesis. The full-term amniotic fluid during delivery is collected either during cesarean section or normal delivery after the artificial rupture membrane procedure.

Earlier, amniotic fluid cells have been routinely cultured from first and mid-term pregnancies for clinical purposes to screen for genetic abnormalities in babies.

Stem Cell Laboratory Techniques. https://doi.org/10.1016/B978-0-12-823729-8.00004-X

Moschidou et al. (2013) were the first to isolate and identify a group of cell population in mid-term pregnancies that has the attributes of broadly multipotent cells and potential for therapeutic purposes. Due to the multipotent differentiation capabilities, most researchers used amniotic fluid stem cells (AFSCs) or amniotic fluid mesenchymal stem cells (AF-MSC) to describe stem cells isolated from amniotic fluid (AF). There are several protocols established for the isolation and differentiation of AFSCs. The isolation step in isolating stem cells from the amniotic fluid of the one-stage culture was first introduced by Prusa and Hengstschlager (2002). Later in 2004, Tsai (2004) developed a two-stage culture to achieve more robust AF-MSCs. An additional step of isolating stem cells based on immunoselection against c-kit$^+$ was introduced and has been shown to enrich a population of the highly potent stem cells, known as AFSCs (de Coppi et al., 2007).

Alternatively, AFSCs could also be derived from the full-term AF harvested from either the cesarean section procedure (C-sect) or spontaneous vaginal delivery (reviewed in Hamid et al., 2017). AFSCs from these sources offer an alternative in isolating multipotent cells as the amniotic fluid from full-term pregnancy is often discarded as biological waste. In addition, sample collection does not pose any known risk to the patient. The first discovery of highly potent AFSCs from full-term pregnancies was demonstrated by You et al. (2008) using a one-stage protocol in which the AFSCs isolated using this method were shown to express Oct-4, a stem cell marker, and these cells could differentiate into osteogenic. Following the discovery, more studies were done to isolate AFSCs from full-term human pregnancies using the one-stage protocol (Bottai et al., 2012; Dolin et al., 2018; Moraghebi et al., 2017; Spitzhorn et al., 2017; Vadasz et al., 2014) and c-kit isolation (Schiavo et al., 2015).

This chapter describes how AFSCs are isolated from the human amniotic fluid at full-term gestation. The fluid is collected during cesarean-section delivery (Fig. 1A–E) and cultured as primary culture until confluent. Isolation of the stem cells can be done through immunoselection against the stem cell receptor marker, c-kit (CD117), or the one-stage culture by splitting the confluent cells cultured in the same medium as the primary culture (Fig. 1, III). Once confluent, the stem cells are characterized at cellular and molecular levels.

2. Materials, reagents, chemicals, apparatus, and kits

2.1 Chemicals

- 0.1 mM β-mercaptoethanol (Cat No: 21985023; Gibco; New York, United States)
- 0.1% gelatin (Cat No: G2500; Sigma-Aldrich; New York, United States)
- 0.25% trypsin-EDTA (Cat No: 15050057; Gibco; New York, United States)
- 0.3% bovine serum albumin (BSA) (Cat No: PC0903; New York, United States)
- 0.4% trypan blue (Cat No: T8154; Sigma-Aldrich; New York, United States)
- 0.5 mM sodium pyruvate (Cat No: 11360070; Gibco; China)
- 1 × nonessential amino acids (Cat No: 11140050; Gibco; Paiesley; UK)
- 1 × phosphate-buffered saline (PBS) (Cat No: 70011044; Gibco; New York, United States)

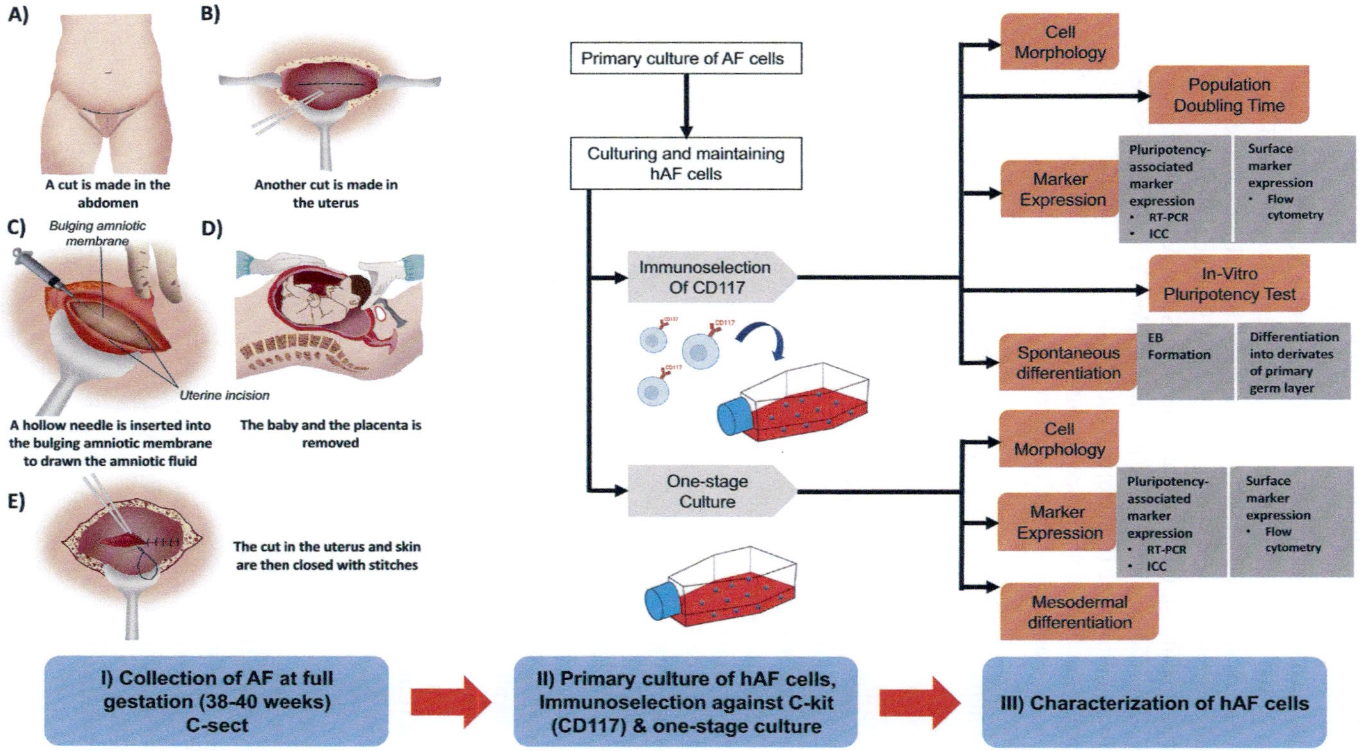

FIG. 1

Overall flowchart describing the establishment and characterization of human full-term AF cells. (I) Procedure involved during collection of the AF at full gestation (38–40 weeks of pregnancy) via C-sect. (II) Primary culture of hAF cells and isolation of full-term hAFSCs via immunoselection against c-kit (CD117) and one-stage culture. (III) Characterization of hAF cells after immunoselection of CD117 and one-stage culture.

- 1 × tris-acetate buffer (Cat No: PB0940; New York, United States)
- 1% L-glutamine (Cat No: 25030081; Gibco, Brazil)
- 1% penicillin-streptomycin (Cat No: 15140122; Gibco; New York, United States)
- 1% Triton X-100 (Cat No: X100; Sigma-Aldrich; St. Louis, United States)
- 10% formalin (Cat No: HT501128; Sigma-Aldrich; St. Louis, United States)
- 15% fetal bovine serum (FBS) (Cat No: 26140079; Gibco, Brazil)
- 18% Chang B (Cat No: C100; Irvine Scientific; Santa Ana, United States)
- 2.0mM EDTA (Cat No: 15575020; Invitrogen; New York, United States)
- 2% antibiotic-antimycotic (Cat No: 15240062; Gibco; New York, United States)
- 2% Chang C (Cat No: C106; Irvine Scientific; Santa Ana, United States)
- 2.3% sodium bicarbonate ($NaHCO_3$) (Cat No: 25080094; Gibco; New York, United States)
- 2.0mM sodium hydroxide (NaOH) (Cat No: S3881; Sigma-Aldrich; St. Louis, United States)
- 4% paraformaldehyde (PFA) (Cat No: 158127; Sigma-Aldrich; Steinheim, Germany)
- 5 μg/mL 4′,6-diamidino-2-phenylindole (DAPI) (Cat No: D1306; Invitrogen; Oregon, United States)
- 60% isopropanol (Cat No: PR315-50; Systerm; Selangor, Malaysia)
- 70% ethanol (Cat No: ET130-50; Systerm; Selangor, Malaysia)
- 98%–100% acetic acid (Cat No: 1000631011; Merck; Darmstadt, Germany)
- agarose gel powder (Cat No: PC0701; Vivantis; New York, United States)
- Alcian Blue staining solution (Cat No: A3157; Sigma-Aldrich; St. Louis, United States)
- Alizarin Red S staining solution (Cat No: A5533; Sigma-Aldrich; St. Louis, United States)
- Amniomax (Cat No: 11269016; Gibco; New York, United States)
- bovine serum albumin (BSA) powder (Cat No: PC0903; New York, United States)
- Glasgow Minimum Essential Medium (GMEM) (Cat No: 11710035, Gibco; New York, United States)
- goat serum (Cat No: 16210064; Gibco; Auckland, New Zealand)
- hAFSCs freezing media (Cat No: M36363-63FM; Celprogen; Torrence, United States)
- iced cold methanol (Cat No: ME313-50; Systerm; Selangor, Malaysia)
- Oil Red O working solution (Cat No: 01391; Sigma-Aldrich; St. Louis, United States)
- secondary antibodies conjugated with IgG-FITC
- sterile distilled water
- Tween 20 (Cat No: P2287; Sigma-Aldrich; St. Louis, United States)
- ultrapure ethidium bromide (EtBr) (Cat No: 15585011; Invitrogen; Oregon, United States)
- α-MEM medium (Cat No: 12571063, Gibco; New York, United States)

2.2 Kits

1. Immunoselection kit (Cat No: 130090312; Miltenyi Biotec; New York, United States)

- Magnetic assorted cell sorting kit (MACs)
- 30.0 μm nylon mesh or Pre-Separation Filters
- FcR Blocking Reagent
- CD117 MicroBeads
- MACS separator

2. RNA extraction kit (Cat No: QIAG-74104; Qiagen; Hilden, Germany)
   - Lysis buffer
   - 2.0 mL collection tube
   - Genomic DNA (gDNA) eliminator spin column
   - Spin column
   - Wash buffer
   - RNase-free water
3. Adipogenesis differentiation kit (Cat No: A1007001; Gibco; California, United States)
4. Chondrogenesis differentiation kit (Cat No: A1007101; Gibco; California, United States)
5. Osteogenesis differentiation kit (Cat No: A1007201; Gibco; California, United States)
6. β-Galactosidase staining kit (Cat No: K802; BioVision; California, United States)
   - Staining solution mix (staining solution, staining supplement, X-gal in DMF)

2.3 Solution and reagent preparation

1. 0.1% gelatin (100.0 mL)	
Gelatin from porcine skin, Type A	0.1 g
1 × PBS	100.0 mL
2. 4% paraformaldehyde (PFA) (250.0 mL)	
95.0% PFA	10.0 g
2.0 mM sodium hydroxide (NaOH)	6.25 mL
1.0 × PBS	To 250.0 mL
Note: Dissolve PFA in a water bath with temperature of 55.0°C and adjust pH to 7.2 by adding HCL (Cat No; 320331; Sigma-Aldrich; St. Louis, United States).	
3. Alcian Blue staining solution (100.0 mL)	
Alcian Blue 8 GX	0.1 g
98%–100% ethanol	60.0 mL
98%–100% acetic acid	40.0 mL
4. Alizarin Red S staining solution (100.0 mL)	
Alizarin Red S	2.0 g
Sterile distilled water	100.0 mL
Note: Adjust the pH to 4.1–4.3 with HCL or NH_4OH. Filter the dark-brown solution and store in the dark before use.	

Continued

5. Blocking solution (100.0 mL)	
Bovine serum albumin (BSA) powder	0.3 g
Tween 20	1.0 mL
Goat serum	1.0 mL
1 × PBS	To 100.0 mL
6. Chang medium (250.0 mL)	
18% Chang B	45.0 mL
2% Chang C	5.0 mL
1% Penicillin-streptomycin	2.5 mL
1% L-glutamine	2.5 mL
15% fetal bovine serum (FBS)	37.5 mL
α-MEM medium	157.5 mL
7. Degas buffer	
1 × PBS (pH: 7.2)	X μL
0.5% fetal bovine serum (FBS)	X μL
2.0 mM EDTA	X μL
Note: "X" is referring to the volume need to be added to achieve the final concentration of the respective component.	
8. Embryoid body medium (EBM) (500.0 mL)	
Glasgow Minimum Essential Medium (GMEM)	403.05 mL
1.0 mM L-glutamine	2.5 mL
0.5 mM sodium pyruvate	2.5 mL
2.3% sodium bicarbonate	11.5 mL
1 × Nonessential amino acids	5.0 mL
0.1 mM β-mercaptoethanol	0.45 mL
15% fetal bovine serum (FBS)	75.0 mL
9. Oil Red O staining solution (100.0 mL)	
Oil Red O powder	0.3 g
99% isopropanol	100.0 mL
Note: Mix three parts of Oil Red O stock solution with two parts of sterile distilled water.	
10. Permeabilization solution (55.0 mL)	
1.0% Triton X-100	5.0 mL
1.0 × PBS	50.0 mL
Note: To prepare 1% Triton X-100 (Cat No: X100; Sigma-Aldrich; St. Louis, United States), dilute 1% Triton X-100 (Cat No: X100; Sigma-Aldrich; St. Louis, United States) with 0.1% 1 × PBS.	
11. Reverse transcription mix (10.0 μL)	
5.0 × Reaction buffer	4.0 μL
7.5 mM $MgCl_2$	3.0 μL
0.5 mM PCR nucleotide mix (each dNTP)	1.0 μL
Recombinant RNasin ribonuclear inhibitor	1.0 μL
Reverse transcriptase	1.0 μL

Continued

12. RNA and primer mix (10.0 μL)	
Experimental RNA (500.0 ng–1.0 μg)	X μL
Oligo(dT)$_{15}$ Primer (0.5 μg/reaction)	1.0 μL
Random Primer (0.5 μg/reaction)	1.0 μL
Nuclease-free water	X μL
Note: "X" is referring to the volume that need to be added to achieve the final concentration of the respective component.	
13. RT-qPCR reaction mix (20.0 μL)	
2 × qPCR Mastermix	10 μL
Primer/probe mix	1.0 μL
Template (150.0–250.0 ng)	5.0 μL
RNAse/DNAse free water	4.0 μL

2.4 Procedure

2.4.1 Collection of the amniotic fluid (AF)

The amniotic fluid is collected by the obstetrics and gynecology specialist after approval from the Institutional Ethics Committee is obtained. The collection should be done aseptically, and blood-free fluid should be aimed. Generally, 5.0–10.0 mL of the AF is sufficient to propagate primary AF cells followed by isolating the one-stage culture or immunoselection against c-kit$^+$ cells procedure.

Generally, the collection of the AF is done following these steps (Fig. 1A–E) during the c-section procedure:

1. Abdominal cavity is entered in the usual way. A cut is made in the abdomen for caesarean section (C-section), and another cut is made in the uterus.
2. The lower segment of the uterus is cut carefully keeping the amniotic membrane intact. A hollow needle is inserted into the bulging amniotic membrane to draw about 5.0–10.0 mL of the amniotic fluid from a low-level section. The drawn AF was transferred to a sterile container.
3. Baby and placenta are delivered.
4. The uterus and abdomen are closed in the usual manner.

2.4.2 Primary culture of amniotic fluid (AF) cells

1. Take the collected AF to the tissue culture laboratory.
2. Withdraw 5.0–10.0 mL of the AF using a 20.0 mL serological pipette.
3. Pool the AF sample in a 50.0 mL centrifuged tube under sterile conditions.
4. Centrifuge the AF sample at 400 × *g* for 10 min. Discard the supernatant fluid.
5. Wash the cell pellet twice with phosphate-buffered saline (PBS). Discard the supernatant and collect the cell pellet.

   Note: This step is done to ensure blood and other contaminants are removed.
6. Resuspend the cell pellets in a 5.0 mL fresh growth medium (Amniomax) with 2% antibiotic-antimycotic and transfer the cell suspension into a 25.0 cm^2 culture flask in. Tilt the flask gently to spread the cells evenly over the growth surface.

7. Cap the flask and place it in an incubator at 37°C with 5% humidified CO_2.
8. After 14–20 days, discard nonadherent cells and cultivate the adherent cells until they reach 80% confluency.
9. Change the culture medium every 3 days.
10. As they reach 80% confluency, detach the cells with 0.05% trypsin-EDTA and replate at 10,000 cells/cm^2 under the same culture conditions (Fig. 2).

2.4.3 Immunoselection of c-kit$^+$ human full-term amniotic fluid stem cells (hAFSCs)

Perform immunoselection of c-kit$^+$ cells using the magnetic assorted cell sorting (MACs) kit (Fig. 3).

1. *Preparation of sample.* (Prepare approximately 1×10^8 hAFSCs1 prior to magnetic labeling).
   a. Yield single-cell suspension of hAFSCs (see protocol: Culturing and maintaining hAFSCs and counting of hAFSCs).
   b. Pass cells through 30.0 μm nylon mesh or Pre-Separation Filters.
   c. Wash the cell pellet with degas buffer once.
   d. Resuspend the cell pellet in 300.0 μL of degas buffer to remove clumps.

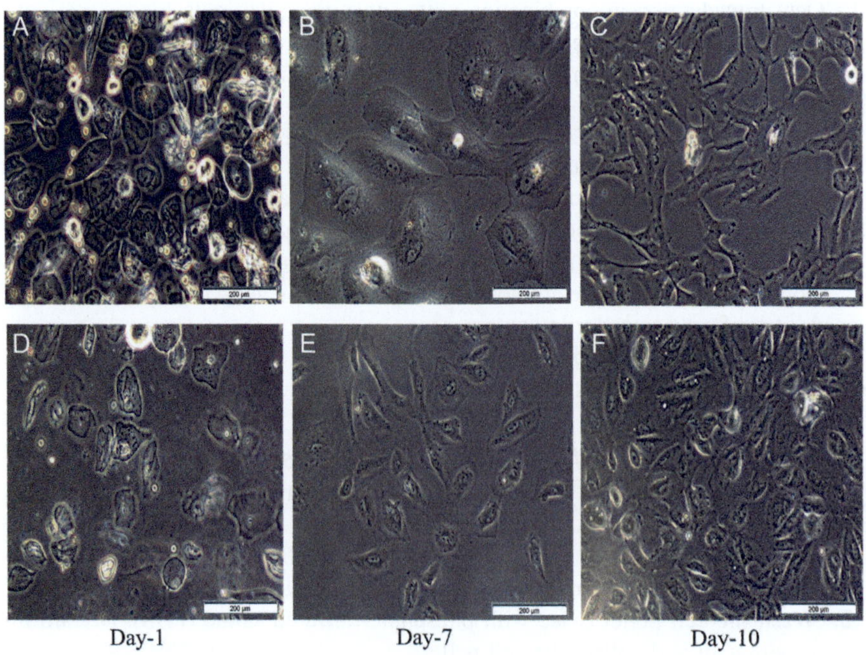

FIG. 2

Primary culture of human AF cells from cesarean-section samples. The cells were in suspension on day-1 and started to attach on day-7 and confluent on day-10. The magnification is ×100.

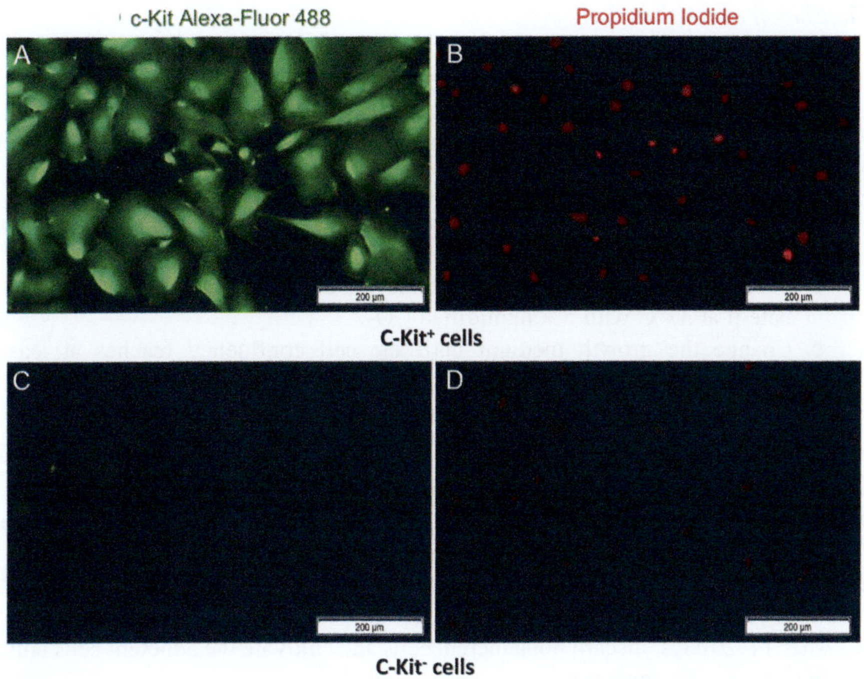

FIG. 3

Expression of c-kit from cesarean-section sample by immunocytostaining revealed high expression of c-kit in c-kit$^+$ cells compared to no expression in c-kit$^-$ cells, indicating the success in isolating the c-kit$^+$ cells (CD117$^+$ cells). The magnification is × 100. Propidium iodide stains the nuclei.

2. *Direct magnetic labeling.*
   a. Add 100.0 μL of FcR Blocking Reagent and 100.0 μL of CD117 microbeads into the cell suspension.
   b. Mix well and incubate for 15 min in a 2–8°C refrigerator.
   c. Wash cells with 1.0 mL of degas buffer.
   d. Centrifuge the cell suspension at 300 × *g* for 10 min. Carefully remove the supernatant without disturbing the cell pellet.
   e. Resuspend the cell pellet in 500.0 μL of degas buffer.
3. *Magnetic separation.* (Ensure proper columns are used when performing separation).
   a. Rinse the MS column with 500.0 μL of degas buffer.
   b. Place the column in the magnetic field of the MACS Separator.
   c. Transfer the cell suspension into the column.
   d. Harvest the flow-through cells and label them as c-kit$^-$ cells.
   e. Wash the column with 500.0 μL of degas buffer and collect the flow-through cells. Repeat this step twice.

4. *Elution of c-kit⁺ cells.*
   a. Take out the column from the separator and put it on an appropriate collection tube.
   b. Add 1.0 mL of degas buffer to the column and immediately push the plunger into the column to flush out the magnetically labeled cells.
5. *Culturing of human full-term amniotic fluid (AF) c-kit$^+$ and c-kit$^-$;cells.*
   a. Dilute the cell suspension for both c-kit$^+$ and c-kit$^-$ cells in 5.0 mL of the Chang medium.
   b. Seed the cells suspension into a prelabeled 25.0 cm^2 culture flask and incubate it at 37°C with 5% humidified CO_2.
   c. Change the growth medium until the cell confluency reaches at least 70.0%–80.0%, indicating the cell may be ready to be subculture (see protocol: Culturing and maintaining hAFSCs and counting of hAFSCs).

2.4.4 One-stage culture of human full-term AFSCs

1. Prepare a single-cell suspension from the pellet obtained from the AF sample (refer primary protocol culture of amniotic fluid (AF) cells).
2. Transfer the cell suspension into a 25.0 cm^2 culture flask and incubate at 37°C with 5% humidified CO_2 for 14–20 days.
3. After 14–20 days, discard nonadherent cells and cultivate the adherent cells until they reach 80% confluency.
4. Change the culture medium every 3 days.
5. As they reach 80% confluency, detach the cells with 0.05% trypsin-EDTA and replate at 10,000 cells/cm^2 under the same culture conditions.
6. Observe the cultured cells under a routine inverted microscope (Fig. 4).

2.4.5 Propagation of full-term human AFSCs

1. *Culturing and maintaining full-term hAFSCs.* Examine the cells carefully under an inverted light microscope for any sign of contamination or deterioration. If subculture is required, proceed as follows:
   a. Clean up the biosafety cabinet by wiping it out with 70% alcohol.
   b. Wipe the material and reagent bottles with 70% alcohol before placing them in the biosafety cabinet.
   c. Take the tissue culture flask into the sterile working area.
   d. Carefully uncap the culture flask.
   e. Withdraw the old growth medium using a sterile serological pipette and discard it into a waste beaker.
   f. Discard the pipette.
   g. Using a sterile serological pipette, add 37°C prewarmed 1 × PBS to the side of the flask. Discard the PBS once rinsed over the cells. Repeat this step twice.
   h. Add 1.0 mL of 0.25% trypsin-EDTA to the side of the flask and spread the liquid to ensure the monolayer is wholly covered. Incubate the tissue culture flask in an incubator at 37°C with 5% humidified CO_2 to enhance the

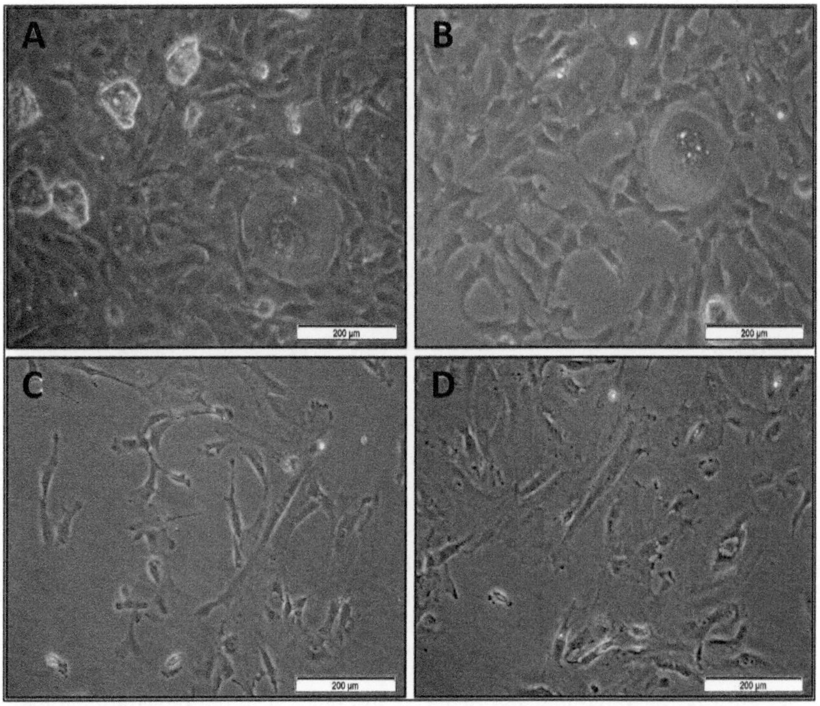

FIG. 4

Heterogeneity of human AF cells. The human AF cell culture was observed to be heterogenous with morphology of AF-type cells (A, B), F-type cells (C), and E-type cells (D). The magnification is × 100.

detachment of cells. This process should take no longer than 5 min as it may lead to cell lysis.

i. Add 1.0 mL of fetal bovine serum (FBS) and disperse the cells by repeated pipetting over the surface of the flask.
j. Gently pipette the cell suspension and transfer it into a 50.0 mL centrifuge tube.
k. Centrifuge the cell suspension at $200 \times g$ for 5 min. Discard the supernatant fluid and collect the cell pellet.
l. Resuspend the cell pellet in a growth medium (Amniomax) and gently pipette the cell suspension up and down to disperse the cells into a single-cell suspension.
m. Count the cells with a hemocytometer (see protocol: Counting of AFSCs) and record the cell counts.
n. Dilute the cells suspension to the appropriate seeding density.
o. Seed the cell suspension into a new 25 cm^2 culture flask at the cell density of 10,000 cells/cm^2 in 5.0 mL of the growth medium (Amniomax) with 2.0% antibiotic-antimycotic.
p. Cap the flask and return them to the incubator.

2. *Counting of full-term hAFSCs.*
   a. Prepare a single-cell suspension (see protocol: Culturing and maintaining hAFSCs).
   b. Take 10.0 μL of the cell suspension and gently mix and pipette up and down with 10.0 μL of 0.4% trypan blue (1 to 1 ratio).
   c. Count the viable cells using a hemocytometer under the inverted light microscope.
      Note: Count the cells within 5 min *to prevent cell death.*
   d. Record the cell counts and calculate the number of viable cells using the following formula:

$$\text{Viable cell/mL} = \frac{\text{Total cell counted on the grid}}{4} \times 10^4 \times \text{Volume of cell suspension} \times \text{Dilution factor.}$$

3. *Cryopreservation of full-term hAFSCs.*
   a. Grow the cells until they reach 80% confluency.
   b. Trypsinize and count the cells (see protocol: Culturing and maintaining hAFSCs and counting of hAFSCs).
   c. Resuspend the pellet in 1.0 mL of hAFSCs freezing media.
   d. Transfer the cell suspension into prelabeled cryovial and cap the cryovial.
   e. Place the cryovial into a −80°C freezer.

4. *Thawing frozen full-term hAFSCs.*
   a. Prepare the apparatus and material and place it in the biosafety cabinet.
   b. Take out the cryovial from the −80°C freezer and ensure the label before proceeding.
   c. Quickly thaw the cryovial by gently swirling it in a 37°C water bath until just a small bit of ice is left in the vial.
   d. Wipe the cryovial with 70% alcohol and transfer it into the biosafety cabinet.
   e. Resuspend the thawed cells with 1.0 mL of the growth medium (Amniomax, Gibco-Invitrogen).
   f. Transfer the cell suspension into a 15.0 mL centrifuge tube using a 1.0 mL pipette.
   g. Centrifuge the cell suspension at 200 × *g* for 5 min.
   h. Carefully remove the supernatant medium without disturbing the cell pellet.
   i. Resuspend the cell pellet in 5.0 mL of the growth medium (Amniomax) with 2% antibiotic-antimycotic.
   j. Seed the cell suspension in the 25.0 cm^2 culture flask and incubate them at 37°C with 5% humidified CO_2.

3. Characterization of c-kit cells/AFSCs

3.1 Cell morphology

Observation of the cells via microscopic analysis by using an inverted light microscope.

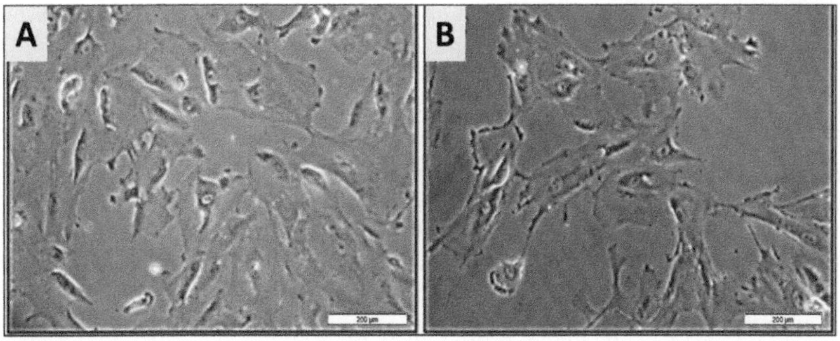

FIG. 5

Morphology of c-kit$^+$ cells in the culture. Human c-kit$^+$ cells show more of AF-type cells (A, B). The magnification is $\times 100$.

1. Clean up the microscope stage with 70% alcohol, and wipe the objective lens with lens tissue, if necessary.
2. Put the $25.0\,cm^2$ culture flask on the microscope stage.
3. Begin to examine the cells with low magnifications (using $4\times$ and $10\times$ phase-contrast objectives), followed with high magnifications (using $20\times$ and $40\times$ phase-contrast objectives).
4. Monitor the growth phase, such as sparse, confluency, density, etc.
5. Observe the state of the cells such as morphology, granularity, vacuolated, etc.
6. Check the condition of the growth medium for signs of contamination, pH changes, and presence of cell debris.
7. Record your observations and snap images of the cells for future references (Fig. 5).
8. Return the flask to the incubator.

3.2 Population doubling time

For the calculation of the cell doubling time, culture the cells from Passage 1 until Passage 6.

1. Culture the cells in the $25.0\,cm^2$ tissue culture flask at initial cell density of $10{,}000\,cells/cm^2$ in 5.0 mL of the growth medium (Amniomax) with 2% antibiotic-antimycotic.
2. Trypsinize the cells as they reach 70%–80% of cell confluency (see protocol: Culturing and maintaining hAFSCs and counting of hAFSCs).
3. Mix the suspension of cells with trypan blue and count the cells using a hemacytometer (see protocol: Counting of hAFSCs).
4. Record the total number of viable cells present. Repeat each cell passage for at least duplicates.

5. Calculate the mean population doubling time of hAFSCs using an online formula (Retrieved from: www.doubling-time.com).

3.3 Growth kinetic analysis

For the analysis of the cell growth kinetic, culture the cells up until Passage 3.

1. Trypsinize the cells and prepare the single-cell suspension as they reach 70%–80% of cell confluency (see protocol: Culturing and maintaining hAFSCs and counting of hAFSCs).
2. Count the cells with a hemocytometer (see protocol: Counting of AFSCs) and record the cell counts.
3. Dilute the cell suspension to the appropriate seeding density.
   Seed the cell suspension into a 24-well plate at the cell density of 1000 cells/cm^2 in 500.0 μL of growth medium (Amniomax) with 2% antibiotic-antimycotic.
4. Detach the cells with treatment of 500 μL of 0.25% trypsin-EDTA (see protocol: Culturing and maintaining hAFSCs and counting of hAFSCs).
5. Count the cells with a hemocytometer (see protocol: Counting of AFSCs) at day-1 to day-7 and record the cell counts.
6. Perform the experiment in triplicate.

3.4 RNA and protein expression

Characterize the characteristic of c-kit$^+$ cells with *OCT-4*, *NANOG*, and *SOX2* markers for pluripotency. Perform RNA analysis, cDNA synthesis, and reverse transcriptase polymerase chain reaction (RT-PCR) to access the gene expression of pluripotent cells. Then, carry out immunostaining and flow cytometry to evaluate the protein expression of pluripotent cells.

1. *Gene expression*
   (a) **RNA extraction.** Extract RNA according to the protocol recommended by the manufacturer.
      i. Prepare a single-cell suspension of hAFSCs to reach 70%–80% cell confluency (see protocol: Culturing and maintaining hAFSCs and counting of hAFSCs).
      ii. Add 350 μL lysis buffer into the cell pellet and a 2.0 mL collection tube.
      iii. Vortex the tube for 30 s.
      iv. Transfer the content into the genomic DNA (gDNA) eliminator spin column.
      v. Centrifuge the tube at 8000 × *g* for 30 s.
      vi. Discard the column, and the flow-through is collected in the collection tube.
      vii. Add 350.0 μL of 70% ethanol to the flow-through and mix it via pipetting.

viii. Transfer the content into the spin column placed in a 2.0 mL collection tube.

ix. Close the lid gently and centrifuge the tube at $\geq$8000 × *g* for 15 s. Discard the flow-through after centrifugation.

x. Add 700.0 μL buffer to the RNeasy spin column and centrifuge the tube for 15 s at $\geq$8000 × *g*. Discard the flow-through.

xi. Add 500.0 μL of washing buffer to the spin column and centrifuge the tube at $\geq$8000 × *g* for 2 min. Discard the flow-through. Repeat this step twice. *Note: This long centrifugation step is to ensure the spin column membrane is dried enough so that none of the ethanol is carried over. Beware that ethanol contamination may interfere with downstream reactions.*

xii. Place the spin column in a new 1.5 mL collection tube. Add 30.0 μL RNase-free water directly to the column membrane.

xiii. Centrifuge the collection tube for 1 min at 8000 × *g* to elute the RNA.

xiv. Determine the concentration and quality of yielded RNA using Nanodrop and gel electrophoresis.

(b) Agarose gel electrophoresis of RNA and PCR product

i. Heat 0.4 g agarose gel powder in 40.0 mL 1 × Tris-Acetate Buffer (TAE Buffer) using a microwave until it dissolves.

ii. Leave the mixture to cool to 60°C.

iii. Add 1.0 μL Ultrapure Ethidium Bromide (EtBr) into the dissolved agarose mixture.

iv. Pour the agarose mixture into a gel cast and leave it to harden for about 20–30 min at room temperature.

v. Add 1 × loading dye to 1.0 μg genomic material and load it into a gel. For RNA, electrophorese it at 60.0 V for 60.0 min; as for PCR, electrophorese it at 90.0 V for 45 min. Use 50 bp DNA ladder as reference.

vi. Visualize the band produced under ultraviolet (UV) light.

(c) Complementary DNA (cDNA) synthesis. Synthesize first-strand using the cDNA synthesis kit.

i. Mix the RNA and primer mix in a 1.5 mL centrifuge tube and briefly centrifuge the tube.

ii. Heat the tube in a thermal cycler at 70°C for 5 min and immediately chill it on ice for 5 min.

iii. Combine 10.0 μL RNA and primer mix with 10.0 μL of reverse transcription mix.

iv. Incubate the tube at 25°C for 5 min for annealing and 42°C for 1 h for extension.

v. Store the reaction mixture at −20°C before proceeding with conventional PCR and RT-qPCR.

(d) Reverse transcriptase polymerase chain reaction (RT-PCR)

i. Transfer the reaction mixture from −20°C into the biosafety cabinet.

ii. Aliquot 20.0 μL of the reaction mixture and place the samples onto a thermocycler.
iii. Carry out the amplification of synthesized cDNA for 28–35 min at 95°C (1 min), 50–60°C (1 min), 72°C (1 min), and 72°C (5 min).
iv. Once finished, proceed with gel electrophoresis (see protocol: Agarose gel electrophoresis of RNA and PCR product, Fig. 6).

2. *Protein expression*
   (a) **β-Galactosidase staining.** Stain the cells according to the protocol made by the manufacturer. Choose cells at Passage 2 and Passage 6. Culture the cells in a 24-well plate.
      i. Prepare the material and apparatus and place them into the biosafety cabinet.
      ii. Take the 24-well plate and place it in the sterile biosafety cabinet.
      iii. Remove the old growth medium carefully without disturbing the growth area of the cells.
      iv. Gently wash the cells using 500.0 μL of 1 × PBS.
      v. Fix the cells with 500.0 μL of fixative solutions for 15 min at room temperature.
      vi. Discard the fixative solutions and wash the cells with 500.0 μL 1 × PBS. Repeat this step twice.
      vii. Add 500.0 μL of the staining solution mix (staining solution, staining supplement, X-gal in DMF).

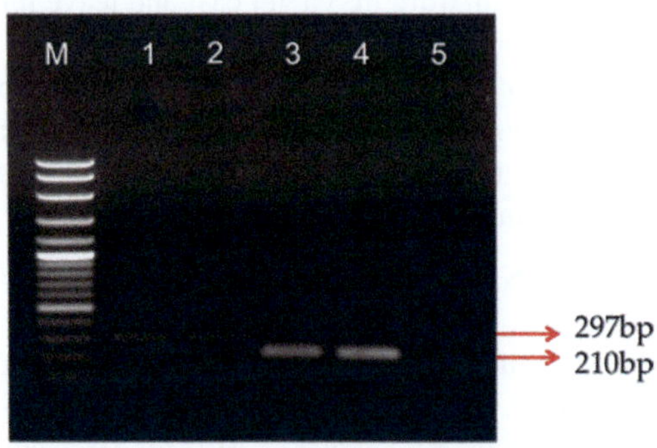

FIG. 6

Expression of pluripotency marker, *Oct-4* on human c-kit$^+$ cells by RT-PCR. Lane 1 shows the expected band size of *Oct-4* expression for positive control; mesenchymal stem cells (MSC) (297 bp). Lane 2 indicates the *Oct-4* expression of human c-kit$^+$ cells (297 bp). Lanes 3 and 4 indicate the internal control, GAPDH expression for human MSC (positive control) and human c-kit$^+$ cells, respectively (210 bp). Lane 5 is the negative control. M indicates the 100 bp marker.

viii. Cover the plate with aluminum foil and incubate it overnight at 37°C.
ix. Observe the cells under the inverted microscope for the development of blue colors that indicate senescence cells.

(b) Immunocytochemistry (ICC) staining of hAFSCs. Ensure to culture hAFSCs in a 24-well plate.

i. Observe the cells under the inverted microscope for cell confluency. Proceed to the next step if 80% of cell confluency is achieved.
ii. Prepare the material and apparatus and place them inside the biosafety cabinet.
iii. Discard the growth medium gently without disturbing the cells.
iv. Wash the cells gently with 500.0 μL of 1 × PBS. Repeat this step twice.
v. Fixation stage: Fix the cell by adding 500 μL of 4% paraformaldehyde (PFA) for 30 min. Remove the 4% PFA and wash the cells twice with 500.0 μL of 1 × PBS.

vi. Permeabilization stage:
  1. *Intracellular staining*: Incubate the fixed cells with the permeabilization solution containing 1% Triton X-100 for 2 h at room temperature. Remove the solution and wash the cells three times with 500.0 μL of 1 × PBS.
  2. *Surface marker staining*: Skip these steps.

vii. Blocking stage:
  1. *Intracellular staining*: Incubate the cells with the blocking solution for 2 h at room temperature. Remove the solution and wash it with 1 × PBS. Repeat this step twice.
  2. *Surface marker staining*: Skip these steps.

viii. Indirect immunofluorescence: For both intracellular and surface marker staining:
  1. Add primary antibody and incubate at 4°C overnight.
  2. Remove the solution and wash cells with 1 × PBS. Repeat this step twice.
  3. Incubate cells with secondary antibody at room temperature for 2 h in the dark.
  4. Wash the cells with 1 × PBS. Repeat this step twice.
  5. Counterstain the nuclei with 5.0 μg/mL of 4′,6-diamidino-2-phenylindole (DAPI) for 10 min and wash three times with 1 × PBS.
  6. Store the cells at 4°C in the dark until they are ready to view.
  7. View the cells under an inverted fluorescence microscope (Olympus). Save the images for further analysis via the Olympus cellSens Standard software (Fig. 7).

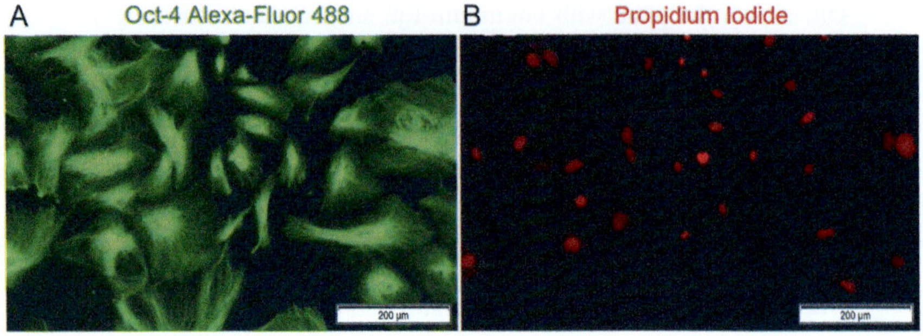

FIG. 7

Oct-4 expression of human c-kit$^+$ cells by ICC.

(c) Fluorescent-activated cell sorting (FACS) flow cytometry analysis.

i. Grow hAFSCs up to 80% confluency. Detach the cells via trypsinization (see protocol: Culturing and maintaining hAFSCs and counting hAFSCs).

ii. Fixation stage: Fix the cell by adding 4% paraformaldehyde (PFA) for 30 min. Remove the 4% PFA via centrifugation and wash the cells with 1 × PBS twice.

iii. Permeabilization stage: Add iced cold methanol and incubate the cells at 4°C for 40 min. Spin the cells at 200 × *g*, 4°C. Remove the supernatant and wash the cells twice with 1 × PBS.

iv. Indirect immunofluorescence:

1. Incubate the cell with primary antibodies in 0.3% bovine serum albumin (BSA) for 1 h at 4°C.
2. Centrifuge the cells at 200 × *g*, 4°C. Remove the supernatant. Wash the cells with 1 × PBS twice.
3. Incubate the cells with secondary antibodies conjugated with IgG-FITC for 30 min at room temperature.
4. Spin the cells and remove the supernatant. Resuspend the cells in 1 × PBS prior to flow cytometry.

3.5 In vitro differentiation into adipogenic, chondrogenic, and osteogenic lineage

Differentiate the hAFSCs by following the protocol made by the manufacturer for the respective lineage.

1. *Differentiation assay*

   (a) Prepare the apparatus and material and place them in the biosafety cabinet.

   (b) Prepare the differentiation media by mixing 90.0 mL of the respective basal media with 10.0 mL of their corresponding supplements and 1.0 mL of antibiotic-antimycotic.

(c) *Differentiation stage*:
   i. Adipogenic lineage. Conduct the differentiation assay based on the protocol by the StemPro Adipogenesis differentiation kit:
      1. Plate hAFSCs in a 24-well plate at a cell density of 10,000 cells/cm^2 (see protocol: Culturing and maintaining hAFSCs).
      2. Observe the cells and remove the old growth medium to reach 80% cells confluency.
      3. Wash the cells with 1 × PBS twice.
      4. Cultivate the cells with adipogenic differentiation media for 3 weeks.
      5. Assess the cells by evaluating the formation of lipid droplets (see protocol: Oil Red O staining for adipocytes).
   ii. Osteogenic lineage. Perform the differentiation process according to the StemPro Osteogenesis differentiation kit:
      1. Plate hAFSCs in a 24-well plate at a cell density of 5000 cells/cm^2 (see protocol: Culturing and maintaining hAFSCs).
      2. Observe the cells and remove the old growth medium to reach 80% cells confluency.
      3. Wash the cells with 1 × PBS twice.
      4. Culture the cells in osteogenic differentiation media for 3 weeks.
      5. Observe the morphological changes and determine the deposition of the calcified extracellular matrix (ECM) via Alizarin Red staining (see protocol: Alizarin Red S staining for osteoblast).
   iii. Chondrogenic lineage. Differentiate the cells according to the protocol described by the manufacturer:
      1. Cultivate hAFSCs in a 48-well plate at a cell density of 20,000 cells/cm^2.
      2. Once it reaches 80% confluency, remove the old growth medium and wash the cells with 1 × PBS twice.
      3. Incubate cells with chondrogenic differentiation media for 3 weeks.
      4. Assess the differentiation of the cells by observing aggregation of cells and via Alcian Blue staining (see protocol: Alcian Blue staining for chondrocytes).

2. *Differentiation staining*
   (a) Oil Red O staining for adipocytes
   i. Discard the differentiation media and fix the cells with 10% formalin.
   ii. Cover the bottom of each well with 60% isopropanol and incubate for 5 min.
   iii. Remove the isopropanol and add 2.0 mL of the Oil Red O working solution. Incubate the cells for 20 min at room temperature.
   iv. Wash the cells with distilled water until the excess stain is no longer apparent. Repeat this step twice.
   v. Observe the cells under an inverted microscope (Fig. 8A and D).

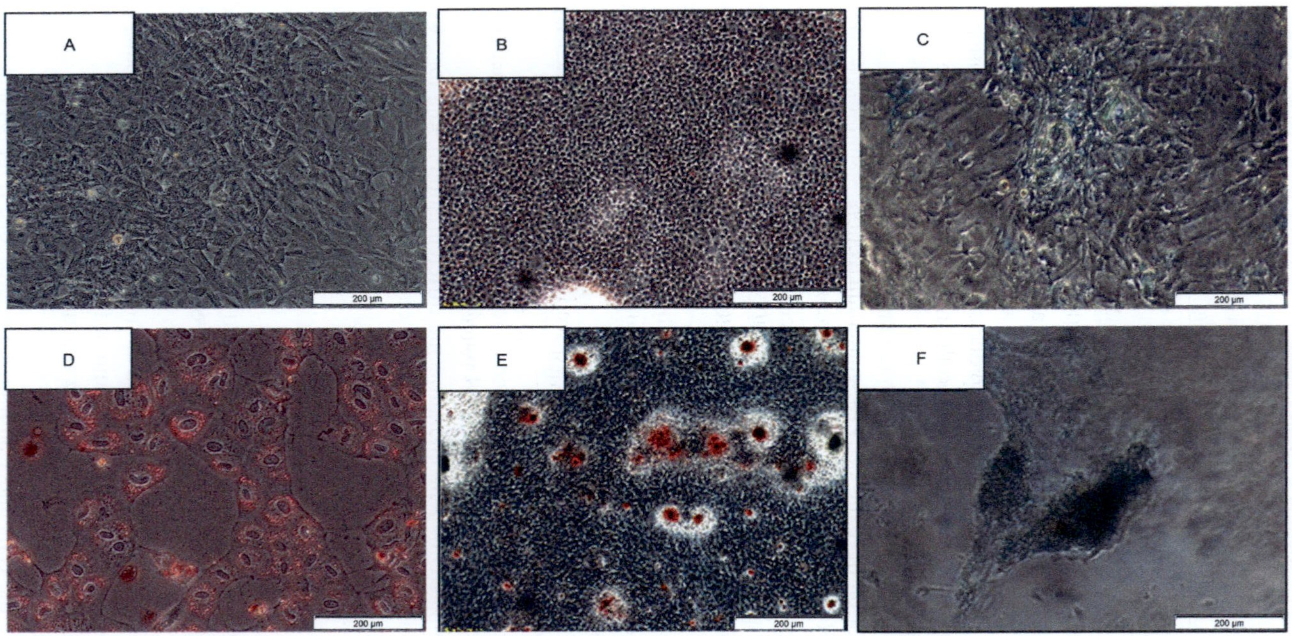

FIG. 8

Monolayer differentiation of full-term AFSCs into adipogenic, osteogenic, and chondrogenic lineage. Undifferentiated full-term AFSCs were negatively stained for Oil Red O (A), Alizarin Red (B), and Alcian Blue (C). Lipid droplets were stained positive by Oil Red O (D). Calcium deposition after monolayer differentiation into osteogenic lineage was confirmed by Alizarin Red staining (E). The induced AFSCs exhibit blue color indicative of the cartilage extracellular matrix by Alcian Blue staining (F).

(b) Alizarin Red S staining for osteoblast

i. Fix cells with 10% formalin. Ensure to discard differentiation media before proceeding.

ii. Wash cells with 1 × PBS twice.

iii. Incubate cells with the Alizarin Red S staining solution for 45 min in the dark at room temperature.

iv. Wash cells four times with distilled water.

v. Kept the cells in 1 × and observe it under an inverted microscope (Fig. 8B and E).

(c) Alcian Blue staining for chondrocytes

i. Fix cells with 10% formalin.

ii. Remove the solution and wash the cells with distilled water.

iii. Incubate the cells with the Alcian Blue staining solution at room temperature for 45 min.

iv. Discard the staining solution and wash the cells with the destaining solution twice gently.

v. Add 1 × PBS to the cells and observe it under an inverted microscope (Fig. 8C and F).

3. *Embryoid bodies (EBs) formation*

(a) Cell suspension method

i. Prepare the apparatus and material and place them in the biosafety cabinet.

ii. Plate 1,000,000 cells in 5.0 mL of EBM in low adherence nonattach Petri dish (see protocol: Culturing and maintaining hAFSCs).

iii. Change the growth medium every 2 days. Using a plastic Pasteur pipette, carefully transfer the suspension of the cells into a clear test tube. Let the EBs get settled at the bottom of the tube for 5 min. Carefully remove the old growth medium and add fresh EBM.

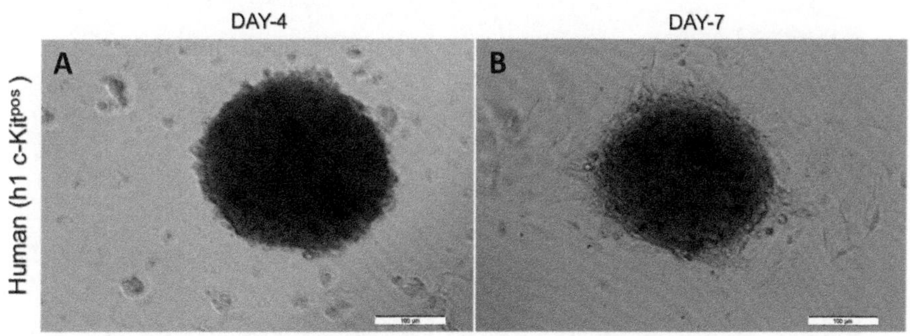

FIG. 9

Formation of embryoid bodies (EB) from human c-kit$^+$ cells on day-4 showing a solid round shape (A) and on day-7, the cells in the EB start to grow and expand (B). The magnification is × 100. Scalebar: 100 μm.

iv. On day-4, transfer the EBs into a 48-well culture plate coated with 0.1% gelatin by using a plastic Pasteur pipette carefully.
v. Observe the EBs under a phase-contrast light inverted microscope.

(b) Hanging drop method

i. Wash the cells twice with 1 × PBS.
ii. Trysinize the cells and harvest the single cells suspension (see protocol: Culturing and maintaining hAFSCs).
iii. Resuspend the cell pellet in the ESC growth medium. Calculate the total number of cells (see protocol: Counting of AFSCs).

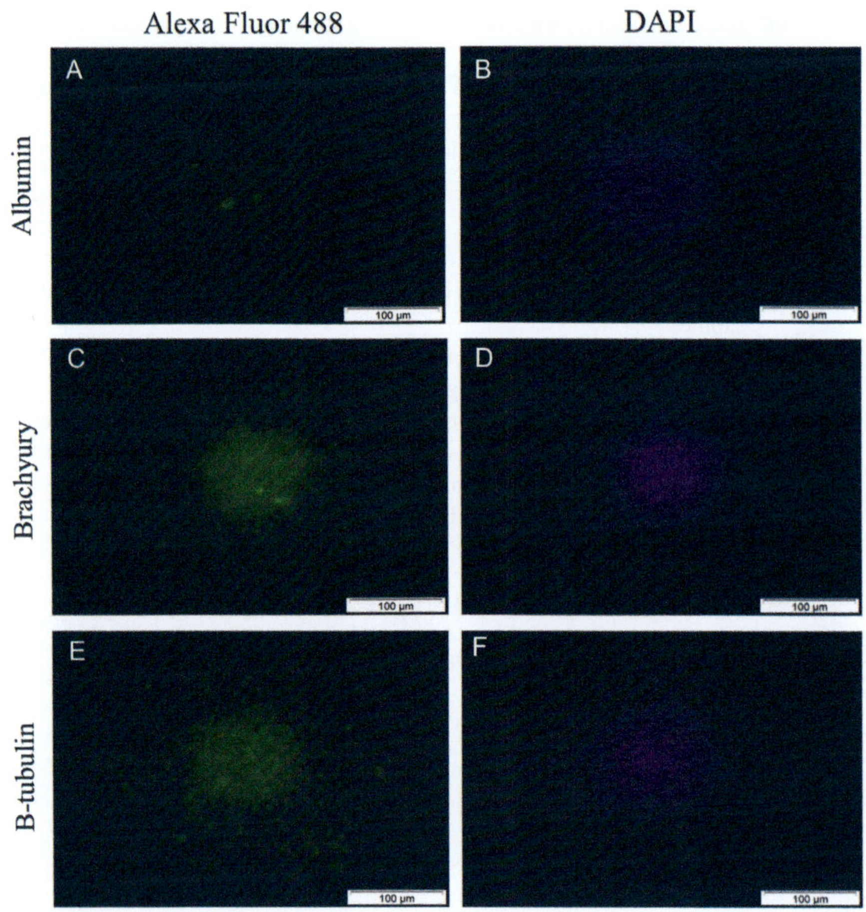

FIG. 10

Embryoid body staining with appropriate markers for endodermal (Albumin), mesodermal (Brachury), and ectodermal (Class III Beta-tubulin) lineages.

iv. Place 20.0 μL of the cell suspension consisting of 4000 cells on the underside of the lid of a Petri dish.
v. Incubate the cells for 2 days.
vi. Transfer the EBs into a Petri dish with the ESC growth medium for 2 days.
vii. Transfer the drops into a 48-well plate coated with 0.1% gelatin using Pasteur pipettes.
viii. View the EBs under a phase-contrast light inverted microscope (Fig. 9).

4. *Spontaneous differentiation of EBs into derivatives of the three primary germ layers*
   a. Pull 10–20 EBs in a gelatin-coated well of a 4-well plate. Let the EBs differentiate and occupy the well for a few days.
   b. For ectodermal derivates such as neurons, stain the EB-differentiated cells with a neuronal marker (such as Tuj1) after 5 days in culture following the immunocytochemistry protocol (Fig. 10).
   c. For mesodermal derivatives, stain the EB-differentiated cells with an appropriate mesodermal marker (such as Brachyury) after 5 days in the culture (Fig. 10).
   d. For endoderm derivatives, stain the EB-differentiated cells with an appropriate endodermal marker (such as GATA4) after at least 18 days in the culture (Fig. 10).

4. Conclusion

In conclusion, human AF cells comprise a heterogeneous cell population isolated from the full-term amniotic fluid, which is often discarded after delivery. There is much evidence that these cells, especially those obtained from mid-term pregnancies, have therapeutic potential. Harvesting cells from the mid-term AF imposes some dangers and should only be done in the case of potentially high-risk pregnancies or genetically affected fetuses. Alternatively, the AF could be obtained from full-term pregnancies delivered via cesarean section and discarded. Previous research has found that full-term AF patients have high-potential stem cells that can be differentiated into adipogenic, osteogenic, chondrogenic, cardiomyogenic, and neurogenic cells. There are evidences showing that AFSCs can be differentiated into cells from three primary germ layers, suggesting the potential of these cells to be used as a cell source for transplantation of degenerative diseases.

References

Bottai, D., Cigognini, D., Nicora, E., Moro, M., Grimoldi, M.G., Adami, R., Abrignani, S., Marconi, A.M., di Giulio, A.M., Gorio, A., 2012. Third trimester amniotic fluid cells with the capacity to develop neural phenotypes and with heterogeneity among sub-populations. Restor. Neurol. Neurosci. 30 (1), 55–68. https://doi.org/10.3233/rnn-2011-0620.

de Coppi, P., Bartsch, G., Siddiqui, M.M., Xu, T., Santos, C.C., Perin, L., Mostoslavsky, G., Serre, A.C., Snyder, E.Y., Yoo, J.J., Furth, M.E., Soker, S., Atala, A., 2007. Isolation of amniotic stem cell lines with potential for therapy. Obstet. Gynecol. Surv. 62 (5), 316–317. https://doi.org/10.1097/01.ogx.0000261640.52511.eb.

Dolin, C.D., Chan, M.K., Basch, R.S., Young, B.K., 2018. Human term amniotic fluid: a novel source of stem cells for regenerative medicine. Am. J. Obstet. Gynecol. 219 (3), 308–309. https://doi.org/10.1016/j.ajog.2018.05.035.

Hamid, A.A., Joharry, M.K., Mun-Fun, H., Hamzah, S.N., Rejali, Z., Yazid, M.N., Karuppiah, T., Nordin, N., 2017. Highly potent stem cells from full-term amniotic fluid: a realistic perspective. Reprod. Biol. 17 (1), 9–18. https://doi.org/10.1016/j.repbio.2017.02.001.

Moraghebi, R., Kirkeby, A., Chaves, P., Rönn, R.E., Sitnicka, E., Parmar, M., Larsson, M., Herbst, A., Woods, N.B., 2017. Term amniotic fluid: an unexploited reserve of mesenchymal stromal cells for reprogramming and potential cell therapy applications. Stem Cell Res. Ther. 8 (1). https://doi.org/10.1186/s13287-017-0582-6.

Moschidou, D., Mukherjee, S., Blundell, M.P., Jones, G.N., Atala, A.J., Thrasher, A.J., Fisk, N.M., de Coppi, P., Guillot, P.V., 2013. Human mid-trimester amniotic fluid stem cells cultured under embryonic stem cell conditions with valproic acid acquire pluripotent characteristics. Stem Cells Dev. 22 (3), 444–458. https://doi.org/10.1089/scd.2012.0267.

Prusa, A.R., Hengstschlager, M., 2002. Amniotic fluid cells and human stem cell research: a new connection. Med. Sci. Monit. 8 (11), RA253–RA257.

Schiavo, A.A., Franzin, C., Albiero, M., Piccoli, M., Spiro, G., Bertin, E., Urbani, L., Visentin, S., Cosmi, E., Fadini, G.P., de Coppi, P., Pozzobon, M., 2015. Endothelial properties of third-trimester amniotic fluid stem cells cultured in hypoxia. Stem Cell Res. Ther. 6 (1). https://doi.org/10.1186/s13287-015-0204-0.

Spitzhorn, L.S., Rahman, M.S., Schwindt, L., Ho, H.T., Wruck, W., Bohndorf, M., Wehrmeyer, S., Ncube, A., Beyer, I., Hagenbeck, C., Balan, P., Fehm, T., Adjaye, J., 2017. Isolation and molecular characterization of amniotic fluid-derived mesenchymal stem cells obtained from caesarean sections. Stem Cells Int. 2017, 1–15. https://doi.org/10.1155/2017/5932706.

Tsai, M.S., 2004. Isolation of human multipotent mesenchymal stem cells from second-trimester amniotic fluid using a novel two-stage culture protocol. Hum. Reprod. 19 (6), 1450–1456. https://doi.org/10.1093/humrep/deh279.

Vadasz, S., Jensen, T., Moncada, C., Girard, E., Zhang, F., Blanchette, A., Finck, C., 2014. Second and third trimester amniotic fluid mesenchymal stem cells can repopulate a decellularized lung scaffold and express lung markers. J. Pediatr. Surg. 49 (11), 1554–1563. https://doi.org/10.1016/j.jpedsurg.2014.04.006.

You, Q., Cai, L., Zheng, J., Tong, X., Zhang, D., Zhang, Y., 2008. Isolation of human mesenchymal stem cells from third-trimester amniotic fluid. Int. J. Gynecol. Obstet. 103 (2), 149–152. https://doi.org/10.1016/j.ijgo.2008.06.012.

Zentelytė, A., Žukauskaitė, D., Jacerytė, I., Borutinskaitė, V.V., Navakauskienė, R., 2021. Small molecule treatments improve differentiation potential of human amniotic fluid stem cells. Front. Bioeng. Biotechnol. 9. https://doi.org/10.3389/fbioe.2021.623886.

CHAPTER

5

Isolation and characterization of c-kit$^+$ stem cells from rat full-term gestation amniotic fluid

Norshariza Nordin[a,b], Siti Sarah Mustaffa Al Bakri[a], Nur Izzati Mansor[a], Devasharmini Mathandaver[a], and Nurfarhana Ferdaos[c]

[a]*Department of Biomedical Science, Faculty of Medicine and Health Sciences, Universiti Putra Malaysia, Serdang, Selangor, Malaysia,* [b]*Genetics & Regenerative Medicine (ReGEN) Research Group, Faculty of Medicine and Health Sciences, Universiti Putra Malaysia, Serdang, Selangor, Malaysia,* [c]*Department of Pharmacology and Chemistry, Faculty of Pharmacy, Universiti Teknologi MARA Puncak Alam Campus, Selangor, Malaysia*

1. Introduction

Different stem cell types with varying stemness have been isolated from humans and animals. To a greater extent, various body compartments and cavities, such as the cardiac cavity, dental pulp, or extraembryonic compartment, contain high potency stem cells. In the Wake Forest Institute of Regenerative Medicine, United States, a group of researchers has successfully isolated cells with therapeutic potential from the mid-term amniotic fluid (AF) of humans and rodents (De Coppi et al., 2007). Their success has put the AF as one of the sources to isolate stem cells with high potency. However, the procurement of the AF at mid-gestation involves an invasive technique, amniocentesis, that could impose certain risks to the mother and the fetus even though the technique is now considered routine. Alternatively, the AF can be obtained from full-term gestation or during delivery.

Full-term and mid-term amniotic fluid stem cells (AFSCs) could be obtained from the AF of different species such as humans, cows, canines, and rodents. The main difference between them is how the AF is obtained for the isolation of AFSCs. For example, procurement of the mid-term AF usually involves amniocentesis, an invasive procedure associated with complications due to amnion sac infection from the needle, sac leakage, and in some cases, miscarriages. In contrast, procurement of the full-term AF is devoid of an invasive issue since it is obtained from the merely discarded AF during normal delivery through the artificial rupture of membrane

Stem Cell Laboratory Techniques. https://doi.org/10.1016/B978-0-12-823729-8.00003-8

(ARM) procedure or cesarean section (Hamid et al., 2017). Besides, in terms of availability of samples, lack of sample accessibility is one of the limitations of the mid-term AF. At the same time, the full-term AF provides an ample amount of AF samples with higher volumes. Mid- and full-term AFSCs of humans and animals have been reported to have similar pluripotency-associated markers, stem cell surface markers, and differentiation potential and be devoid of ethical issues related to their isolation (reviewed in Hamid et al., 2017).

Full-term amniotic fluid stem cells (AFSCs) have been isolated and established from various animals, including mice, rats, bovine, and canine (Hamid et al., 2017). Isolation of AFSCs can be carried out using three methods: (1) one-stage culture, (2) two-stage culture, and (3) immunoselection against c-kit (CD117). The first two methods are more widely used to isolate amniotic fluid mesenchymal stem cells (AF-MSCs), thus explaining the lack of the c-kit expression in these cells. On the other hand, immunoselection against c-kit isolates AFSCs; hence, the cells express c-kit (reviewed in Cananzi et al., 2011). C-kit (CD117) is a type III tyrosine kinase receptor of the stem cell factor (Miettinen and Lasota, 2005). AF-MSCs are categorized as multipotent stem cells as they can differentiate into multiple lineages. On the other hand, AFSCs are characterized as broad multipotent stem cells since the cells can differentiate into derivatives of the three primary germ layers (Mun-Fun et al., 2015) but do not form teratoma when transplanted into a host (Ramasamy et al., 2018).

This chapter describes the procedures to establish rat AFSCs from full-term gestation using immunoselection against c-kit (CD117). Here, the methods to get the amnion sacs from rats, collect the amniotic fluid, culture the amniotic fluid cells, perform immunoselection against c-kit$^+$ cells, and characterize the cells are described in detail. Fig. 1 depicts the steps of establishing AFSCs from the rat full-term amniotic fluid.

2. Materials, reagents, chemicals, apparatus, and kits

2.1 Chemicals

- 2-Mercaptoethanol (Gibco; New York, United States)
- Alcian Blue (Sigma-Aldrich, St. Louis, United States)
- Alizarin Red (Sigma-Aldrich, St. Louis, United States)
- Bovine serum albumin (BSA) 7.5% solution (Gibco; New York, United States)
- Dimethyl sulfoxide ≥99% (DMSO) (Sigma-Aldrich, St. Louis, United States)
- Fetal bovine serum (FBS) (Gibco; New York, United States)
- Gelatin from porcine skin, Type A (Sigma-Aldrich, St. Louis, United States)
- Glasgow's Minimum Essential Media (GMEM) (Gibco; New York, United States)
- Leukemia inhibitory factor rat (LIF3010) (Merck Millipore, United States)
- L-Glutamine (200 mM) (Gibco; New York, United States)
- MEM non-essential amino acid solution (100×) (Gibco; New York, United States)
- Methanol
- Red Oil (Sigma-Aldrich, St. Louis, United States)
- Paraformaldehyde, 95% (PFA) (Sigma-Aldrich, St. Louis, United States)
- Penicillin-Streptomycin (10,000 U/mL) (Gibco; New York, United States)

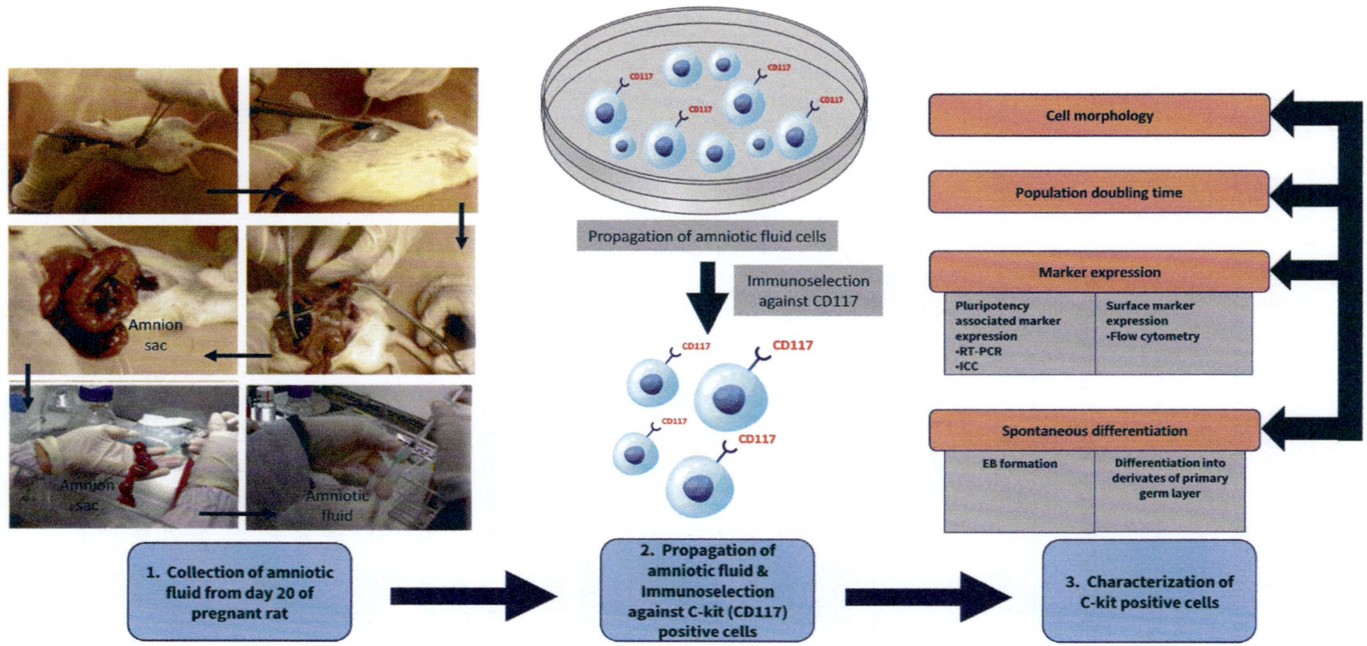

FIG. 1

Generation of AFSCs. The overall steps involved in the establishment of amniotic fluid stem cells (AFSCs) from the rat full-term amniotic fluid.

- Phosphate-buffered saline (10 ×) pH7.4 (PBS) (Gibco; New York, United States)
- Sodium bicarbonate 7.5% solution (Gibco; New York, United States)
- Sodium hydroxide (NaOH) (Sigma-Aldrich, St. Louis, United States)
- Sodium pyruvate (100 mM) (Gibco; New York, United States)
- Tween-20 (Sigma-Aldrich, St. Louis, United States)
- Triton X-100 (Fluka, United States)

2.2 Preparation of solutions and reagents

1.	Stem cell incomplete medium (ESIM)	
	1 × Glasgow's Minimum Essential Media (GMEM)	500.0 mL
	7.5% sodium bicarbonate	2.9 mL
	200.0 mM L-glutamine	2.9 mL
	100.0 mM sodium pyruvate	2.9 mL
	100 × Mem nonessential amino acid	5.9 mL
	55.0 mM 2-mercaptoethanol	1.1 mL
2.	0.1% (w/v) gelatin	
	Gelatin from porcine skin, Type A	0.1 g
	1 × PBS	100.0 mL
3.	Stem cell complete medium (ESM)	
	ESIM	42.5 mL
	100% fetal bovine serum (FBS)	7.5 mL
	10.0 μg/mL rat LIF	50.0 μL
4.	Embryoid body medium (EBM)	
	ESIM	42.5 mL
	100% fetal bovine serum (FBS)	5.0 mL
	100 × Penicillin-Streptomycin	500.0 μL
5.	Fixation solution	
	Paraformaldehyde	4.0 g
	1 × PBS	999.5 mL
	NaOH	500.0 μL
6.	Permeabilization solution	
	100% Triton-X	0.1 mL
	1 × PBS	100.0 mL
7.	Blocking solution	
	Bovine serum albumin	0.3 g
	100 × Goat serum	1.0 mL
	Tween-20	500.0 μL
	1 × PBS	998.5 μL

2.3 Procedure

Before starting to work with the animal tissue, ensure that your work complies with animal ethical rules. Animal ethics approval must be obtained before commencing the animal work.

2.3.1 Collection of amniotic fluid

1. *Induction of estrus in rats.* Put male and female Sprague-Dawley rats together for mating. The timing of successful mating may be determined by examining the vaginas each morning for a hard, mucous plug.
2. *Amniotic fluid collection.* Collect amniotic fluid (AF) samples from time-mated Sprague-Dawley rats ($n=3$) on the 20th day of pregnancy (full-term).
   a. Sacrifice the rats by cervical dislocation and swab the ventral surface with 70% alcohol.
   b. Using sterile forceps and scissors, cut longitudinally along the abdomen midline of the rat, exposing the uterus. At this stage, the uterus containing the fetus is evident in the posterior abdominal cavity (Fig. 2A–C).
   c. Dissect the amnion sacs from the uterus and wash the sacs with ice-cold 1 × PBS (Fig. 2D).
   d. Place the cleaned sacs in a sterile Petri dish in 1 × PBS.
      Note. All preceding steps should be done outside the tissue culture laboratory as animals may carry contamination.
   e. Take the intact amnion sacs to the tissue culture laboratory and transfer them to a fresh and sterile Petri dish (Fig. 2D).
   f. Withdraw the AF from each sac using a sterile 1.0 mL syringe and needle (Fig. 2E).
   g. Pool the AF samples in a 2.0 mL screw-capped vial under sterile conditions (Fig. 2F).
   h. Centrifuge the AF samples at $160 \times g$ for 10 min. Discard the supernatant fluid and collect the cell pellet.
   i. Resuspend the cell pellets in a fresh growth medium (Amniomax, Gibco-Invitrogen) and transfer the cell suspension into a 25.0 cm^2 culture flask. Tilt the flask gently to spread the cells evenly over the growth surface.
   j. Cap the flask and place it in an incubator at 37.0°C, 5% CO_2.
   k. If the cells have adhered, the growth medium volume may be changed regularly over 2–4 days to remove cell debris and unwanted cells until a substantial outgrowth of cells is observed.
   l. Replace the growth medium until the outgrown has spread to occupy at least 70%–80% of the flask area (confluent), at which point the cells may be ready for subculturing (Fig. 3). AF cells were observed to be heterogeneous (Fig. 4).

   Note. Dissection of rats should be performed by qualified personnel such as a certified veterinarian. Do not dissect in the tissue culture laboratory as

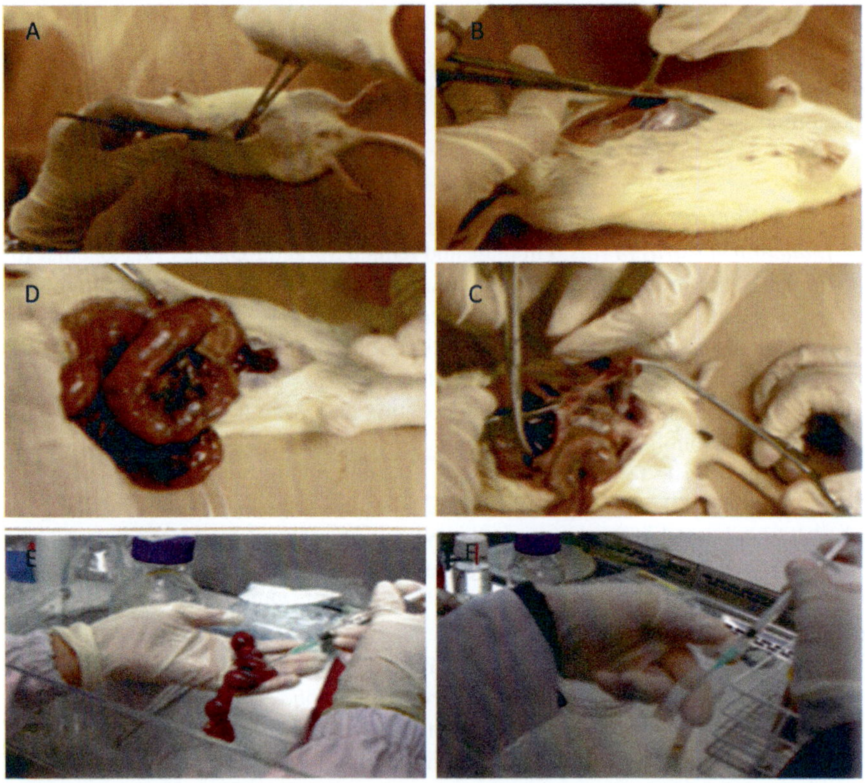

FIG. 2

Collection of amniotic fluid. Collection of the amniotic fluid (AF) from time-mated Sprague-Dawley rats, on the 20th day of pregnancy (full-term). (A) and (B) show a longitudinal cut along the abdomen midline of the rat, exposing the (C) uterus and the dissection of (D) amnion sacs. The sacs were washed with 1 × PBS and brought into the biological safety cabinet (BSC) on ice. Inside the BSC, (E) the AF was withdrawn from each sac using a sterile needle, and (F) pooled into centrifuge tubes.

the animals may spread microbial contamination. Remove the tissue from the animal aseptically, rinse the amnion sac with medium 1 × PBS, and keep the amnion sacs in the PBS. Transfer it to the tissue culture laboratory and transport it as soon as possible in an icebox. If a delay in transferring the sample is inevitable, it can be stored at 4.0°C for 24 h although an immediate transfer provides a better yield. Work with the animal tissue should be then carried out in a Class II biological safety cabinet under sterile conditions.

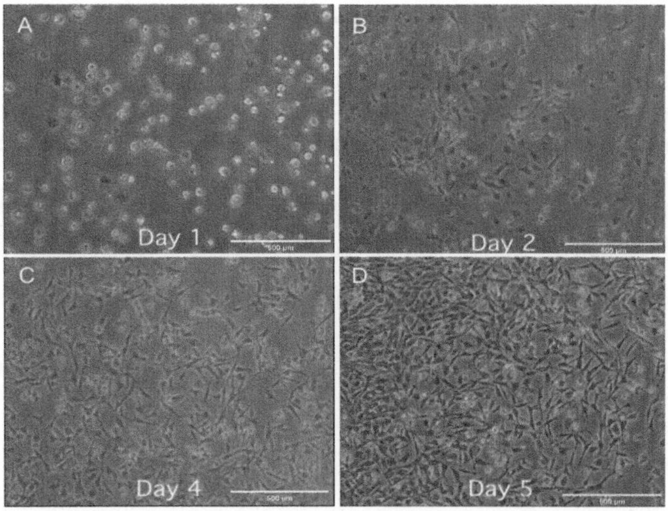

FIG. 3

Primary culture of rat AF cells in Amniomax. On the first day of culture, the cells were still in suspension (A) They started to attach on the second day of culture (B) and continued to grow onward (C) until they were confluent after 5 days in Amniomax (D).

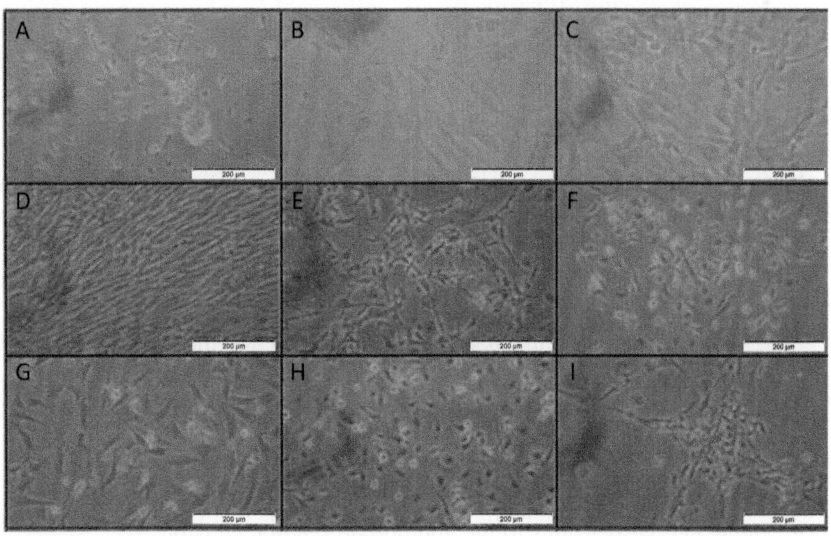

FIG. 4

Morphology of rat AF cells cultured in Amniomax. AF cells grown in Amniomax were observed to be heterogeneous. (A, F) Amniotic fluid (AF)-type cells, (B, C, G) epithelial-like cells (E type), (D, H) fibroblast-like cells (F type), and (E, I) neuronal-like cells were observed.

2.3.2 Propagation of rat full-term amniotic fluid cells

1. *Culturing and maintaining rat AF cells.* Sift through the cells under the inverted light microscope for any sign of contamination or deterioration. If subculture is required, proceed as follows:
   a. Prepare the biological safety cabinet by wiping it with 70% alcohol and ensure whether it is clean.
   b. Bring the reagents and material for the procedures, wipe the reagent bottles and containers with 70% alcohol, and place the items immediately in the biological safety cabinet.
   c. Take the culture flask to the sterile working area.
   d. Uncap the culture flask.
   e. Using a sterile serological pipette, withdraw the old growth medium and discard it into a waste beaker.
   f. Discard the pipette.
   g. Using a sterile serological pipette, add 37.0°C prewarmed 1 × PBS to the side of the flask. (Do not add directly to the cells to avoid dislodging the cells.) Rinse the PBS over the cells and discard. This step is designed to remove the cell debris or traces of serum that would prevent the action of trypsin. Repeat this step twice.
   h. Add 0.25% trypsin with 1 mM EDTA (0.05–0.1 mL/cm^2) to the side of the flask and spread the liquid to ensure that the monolayer is wholly covered. Leave the flask stationary for 15–30 s. If necessary, place it in an incubator at 37.0°C, 5% CO_2 to expedite cell detachment. (Do not leave the flasks longer than 5 min to avoid cell lysis.)
   i. Add a growth medium containing serum (0.1–0.2 mL/cm^2) and disperse the cells by repeated pipetting over the growth surface.
   j. Pipette the cell suspension up and down gently and transfer the cell suspension into a 50.0 mL centrifuge tube.
   k. Centrifuge the cell suspension at 160 × *g* for 5 min. Discard the supernatant fluid and collect the cell pellet.
   l. Resuspend the cell pellet in a growth medium and pipette the cell suspension up and down sufficiently to disperse the cells into a single-cell suspension.

      Note. A single-cell suspension is necessary for the subculture step to ensure an accurate cell count and consistent growth of cells for every passage. It is crucial if quantitative estimates of cell proliferation are being made.

   m. Count the cells with a hemocytometer and record the cell counts.
   n. Dilute cell suspensions to the appropriate seeding density.

- **o.** Seed the cell suspension into a culture flask at the cell density of 1×10^4 cells/cm^2 in a growth medium (0.2 mL/cm^2).
- **p.** Cap the flask and return them to the incubator at 37.0°C, 5% CO_2.

2. *Cell viability study*. Evaluate the optimum condition for the maintenance of rat AF cells.
   - **a.** Prepare a single-cell suspension. (See protocol: Culturing and maintaining rat AF cells.)
   - **b.** Take 10 μL of the cell suspension and mix with 10.0 μL of 0.4% trypan blue (1 to 1 ratio). The number of viable cells was determined using a hemocytometer.

     Note. The unstained cells are identified as viable cells, whereas blue-stained cells are identified as dead cells, as the dye is permeable to damaged cell membranes.
3. *Cryopreservation*.

   Note. Before freezing, check the culture by eye and microscope for (a) morphological characteristics, (b) phase growth cycle, and (c) contamination issues.
   - **a.** Grow the culture until the late log phase.
   - **b.** Trypsinize and count the cells. (See protocol: Culturing and maintaining rat AF cells.)
   - **c.** Resuspend at 2×10^6 cells/mL.
   - **d.** Prepare a freezing medium by adding dimethyl sulfoxide (DMSO) to the 10%–20% growth medium.
   - **e.** Dilute the cell suspension 1:1 with a freezing medium to give approximately 1×10^6 cells/mL and 5%–10% DMSO.
   - **f.** Dispense the cell suspensions into prelabel cryovials and cap the cryovials.
   - **g.** Freeze the cryovials at 1.0°C/min by placing them in a freezing container and placing them at −80°C overnight before transferring them to liquid nitrogen.
4. *Thawing frozen cells*.
   - **a.** Collect all materials, prepare a complete growth medium, and label the culture flask.
   - **b.** Retrieve the cryovials from the freezer and check the label before proceeding.
   - **c.** Thaw the frozen vials by submerging the cryovials in the sterile 37.0°C water bath. (Avoid getting the water up to the cap as this will increase the chance of contamination.)
   - **d.** When the cryovial has thawed, wipe the cryovial with 70% alcohol, and place it in the biological safety cabinet.
   - **e.** Transfer the contents of the cryovial to a 50.0 mL centrifuge tube with a 1 mL pipette.
   - **f.** Dilute the cell suspension to 10.0 mL with a growth medium.
   - **g.** Centrifuge the cell suspension at $160 \times g$ for 5 min.

- **h.** Discard the supernatant medium with the freezing medium.
- **i.** Resuspend the cell pellet in a fresh growth medium.
- **j.** Seed in the culture flask.
- **k.** Check the culture after 24 h. Confirm attachment and estimate the percentage of cell survival under the microscope.

2.3.3 Isolation of c-kit⁺ cells/AFSCs

Perform immunoselection of c-kit$^+$ (CD117) cells using the magnetic assorted cell sorting (MACS) kit (Fig. 5).

1. *Preparation of sample.* Prepare approximately $1 \times 10^6 - 1 \times 10^7$ rat AF cells before magnetic labeling.
   - **a.** Centrifuge the rat AF cells and discard the supernatant medium.
   - **b.** Resuspend the cell pellets in 300.0 μL of degas buffer to make a single-cell suspension. Degas buffer is prepared by adding 1 × phosphate buffered saline (PBS) with pH 7.2 with 0.5% fetal bovine serum (FBS) and 2.0 mM EDTA and transfer the solution into a sterile bottle covered with a rubber stopper with one tube inserted in its hole and connected to a vacuum pump. Turn on the vacuum pump, and as bubbles are observed (due to degas process), swirl the bottle gently to promote the release of dissolved gas. After no bubbles are observed, remove the rubber stopper from the bottle and shut off the vacuum pump.

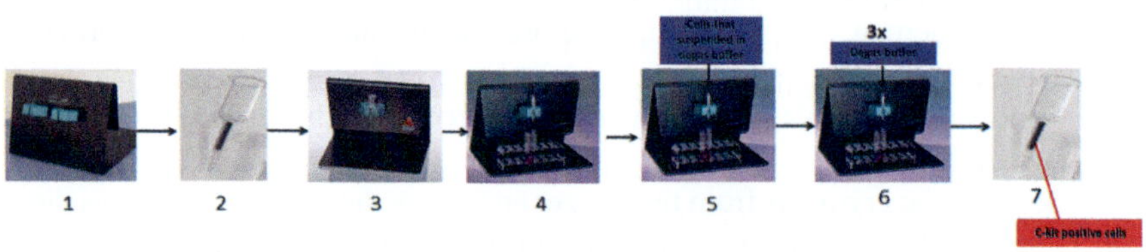

FIG. 5

Isolation of c-kit$^+$ cells using magnetic activated cell sorting (MACS) kit. The sorting process begins with (1) placing the magnet on the board, then (2) rinsing the column with 500.0 μL of degas buffer before (3) subjecting the column to the magnet field. (4) A collection tube for negative cells is placed below the column and (5) the cell suspension (containing the labeled cells) in 500.0 μL of degas buffer is poured into the column, and (6) the column is washed by running through the column with 500.0 μL of degas buffer three times. (7) Finally, the column that contains the c-kit$^+$ cells are taken away from the magnet field and 500.0 μL of degas buffer is added into the column to collect the c-kit$^+$ cells.

2. Direct magnetic labeling.
   a. Add 100 μL of the blocking reagent into a cell suspension and 100.0 μL of c-kit (CD117) microbeads.
   b. Mix well and incubate the mixture at 4–8°C for 15 min.
   c. Before the washing step, add 1.0 mL of buffer and centrifuge the cell suspension at 300 × *g* for 10 min.
   d. Discard the supernatant and resuspend the cell pellet in 500.0 μL of the buffer.
3. *Indirect magnetic labeling*. Optimized the primary c-kit antibody at different concentrations and analyzed using fluorescent activated cell sorting (FACS). Verify the optimum concentration of the primary c-kit antibody. Add the primary c-kit antibody to the cell suspension at the optimum concentration.
   a. Mix well and incubate at 4.0°C for 5 min.
   b. Centrifuge at 300 × *g* for 10 min and remove the supernatant.
   c. Resuspend the cells with 1.0 mL degas buffer to remove the unbound primary c-kit antibody and centrifuge at 300 × *g* for 10 min and remove the supernatant.
   d. Resuspend the cells in 80.0 μL degas buffer.
   e. Add 20.0 μL goat antirabbit IgG microbeads and mix well.
   f. Incubate the cell suspension at 4.0°C for 10 min.
   g. Repeat Steps iii and iv.
   h. Resuspend the cell pellet in 500.0 μL degas buffer.
4. *Magnetic separation*. Perform magnetic separation using appropriate columns.
   a. Rinse the MS Column with 500.0 μL degas buffer and place it in the magnetic field.
   b. Load the labeled cell suspension onto the column.
   c. Collect the cells that passed through and label them as c-kit$^-$ cells
   d. Wash the column by adding degas buffer. Repeat this step twice.
5. *Elution of c-kit$^+$ cells*.
   a. Add 1.0 mL of degas buffer to the column and immediately apply the plunger to flush out the labeled cells (c-kit$^+$ cells).
6. *Culturing of rat AF c-kit$^+$ and c-kit$^-$ cells*.
   a. Precoat the tissue culture flask with sterile porcine gelatin (0.1%, w/v) for at least 20 min. Discard the excess gelatin.
   b. Dilute the cell suspension for both c-kit$^+$ and c-kit$^-$ cells in Chang media on a gelatin-coated tissue culture flask.
   c. Place it in an incubator at 37.0°C, 5% CO_2.
   d. Replace the growth medium until the outgrown has spread to occupy at least 70%–80% confluency, at which point the cells may be ready to be subcultured and stored (see Procedure B). The morphology of c-kit$^+$ cells is shown in Fig. 6.
7. *Enrichment of c-kit$^+$ cells*
   a. Revive cells and grow the cells until 70%–80% confluent.
   b. Subject the c-kit$^+$ cells to undergo a second round of miniMACS to get more homogeneous cells. Repeat the procedure above (Section 2.3.3) for the isolation of c-kit$^+$ cells.

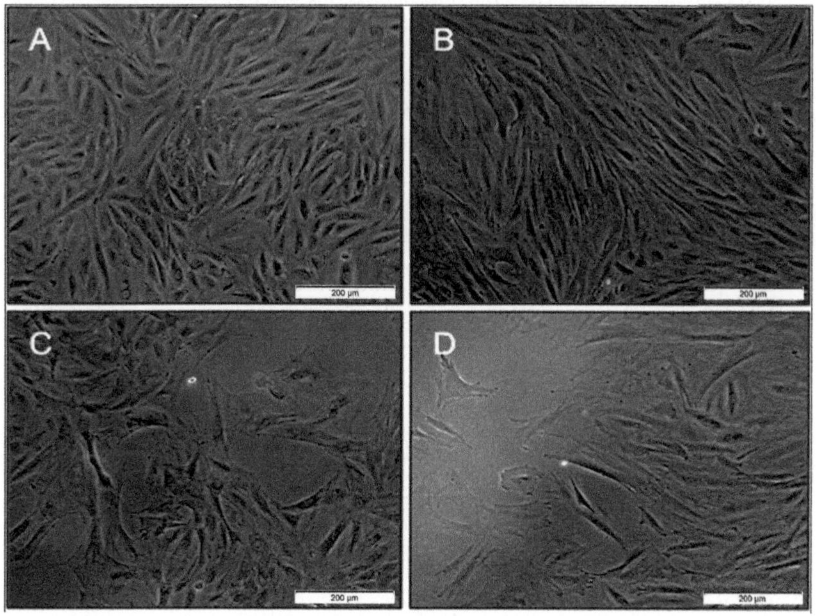

FIG. 6

Morphology of c-kit⁺ cells in Chang medium. Rat c-kit⁺ cells from different samples showing more F-type cells (A, B, C, D). Magnification is 10 ×.

   c. Propagate the enriched c-kit⁺ cells ESM at a cell seeding density of $4 \times 10^4/cm^2$.
   d. Observe the cell morphology under an inverted microscope and capture images for the record.
   e. Passage the cells once confluent or store the cells.

2.3.4 Characterization of c-kit⁺ cells/AFSCs

1. *Cell morphology*. Observation of cells by microscopic analysis—inverted light microscope (Fig. 7).
   a. Wipe the microscope stage with 70% alcohol, and clean the objective lens with lens tissue, if necessary.
   b. Place the tissue culture flasks or dish on the stage.
   c. Examine the cells by starting with low magnifications (using 4 × and 10 × phase-contrast objectives), followed with high magnifications (using 20 × and 40 × phase-contrast objectives).

      Note. For routine examination, a 10 × phase-contrast objective will give adequate detail.
   d. Monitor the growth phase, such as sparse, confluency, density, etc.
   e. Monitor the state of the cells, such as morphology, granularity, vacuolation, etc.

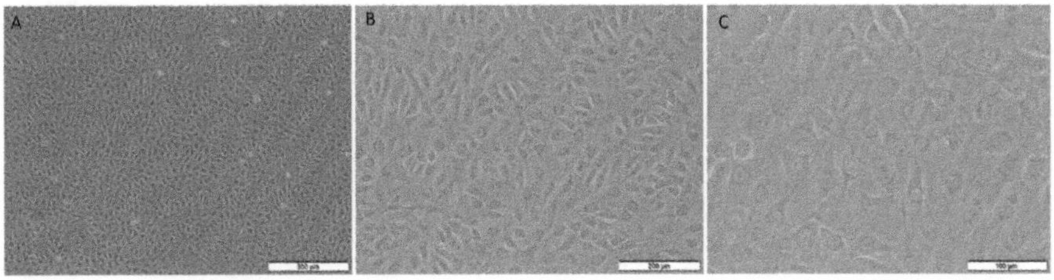

FIG. 7

Morphology of enriched c-kit$^+$ cells cultured in ESM. The morphology of c-kit$^+$ cells viewed under (A) 4 ×, (B) 10 ×, and (C) 20 × magnification showing homogenous cells, displaying a cuboidal AF-type morphology.

   f. Monitor the clarity of the growth medium for signs of contamination, pH changes, and cell debris presence.
   g. Record your observations.
   h. Return the flask to the incubator.
2. *Population doubling time*. To calculate cell doubling time, choose cells randomly from P20–P50.
   a. Grow the cells in the tissue culture flask ($25.0\,cm^2$) in a conventional manner at the initial cell density of 1×10^6 cells.
   b. Harvest cells after 48 h in culture.
   c. Stain cells with trypan blue and count the cells using a hemocytometer. Repeat each cell passage for at least duplicates.
   d. Calculate the mean doubling time using an online formula for PDT (www.doubling-time.com) as in Fig. 8.
   e. The formula for doubling time is as below:

$$\text{Population Doubling Time(PDT)} = \frac{\text{Duration} \times \text{loglog } 2}{\text{loglog(final number of cells)} - \text{loglog(initial number of cells)}}$$

3. *RNA and protein expression*. The characteristics of c-kit$^+$ cells are characterized with Oct-4, Nanog, and Sox2, markers for pluripotent cells.
4. RNA analysis, RT-PCR, is performed to access the gene expression of pluripotent cells; meanwhile, immunostaining and flow cytometry are carried out to evaluate the protein expression of the pluripotent cell markers.
   (a) RNA analysis by RT-PCR. RNA extraction is carried out according to the procedure recommended by the manufacturer.
      i. Grow the cells in the tissue culture flask or dish to give 1×10^6–1×10^7.
      ii. Discard the growth medium. Wash the cells with cold 1 × PBS.
      iii. Harvest the cells. Centrifuge the cells at $100 \times g$–$200 \times g$ for 5–10 min to pellet the cells and discard the supernatant.
      iv. Proceed with RNA extraction protocols.

A

Initial Number of Cells ($\times 10^6$)	Duration (hours)	Final Number of Cells ($\times 10^6$)	PDT
1.00	48	2.50	36.31
1.00	48	2.59	35.84
1.00	48	2.53	37.82
1.00	48	2.38	34.96
1.00	48	2.41	38.37
1.86	48	8.84	21.35
1.69	48	5.60	27.77
0.60	48	2.26	25.09
1.00	48	3.84	24.73
2.58	48	7.84	29.93
2.00	48	5.65	32.04
1.84	48	5.00	33.28
2.00	48	5.00	36.31

B

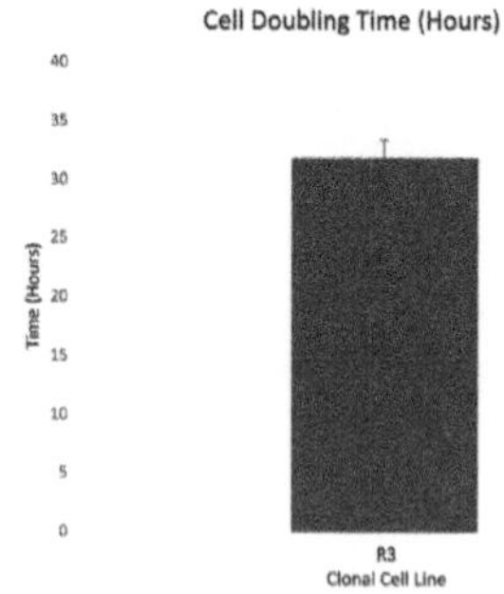

C

Mean	31.83
Standard Deviation (SD)	5.57
Standard Error of Mean (SEM)	1.55

FIG. 8

Calculation of population doubling time (PDT). (A) The initial and final number of cells were counted using a hemacytometer and calculated using the online formula as mentioned in the steps for population doubling time. (B) The average PDT for R3 was 31.83 ± 1.55 h. Results are the representatives of randomly chosen passages (P20–P50) from R3 from more than five independent experiments. (C) The statistical value of the data in Table A.

- **v.** Store it at −70°C before use.
- **vi.** Synthesize cDNA from the RNA according to the manufacturer's instructions.
- **vii.** With reference to the manufacturer's instruction, prepare enough cDNA to amplify the samples.
- **viii.** For the amplification per sample, the reaction mixture must contain a PCR reaction mix, primer set, DNA polymerase, and cDNA.
- **ix.** Aliquot 20.0 μL of the mix into individual PCR tubes.
- **x.** Place the samples onto a thermocycler programmed with the appropriate cycling details as follows:
  - Initial denaturation at 95.0°C (3 min)
  - Amplification of 28–35 cycles at 95.0°C (1 min), 50.0–60.0°C (1 min), and 72.0°C (1 min)
  - Final elongation at 72.0°C (5 min)
  - Hold at 4.0°C until required.
- **xi.** Proceed with gel electrophoresis (Fig. 9).

(b) Immunostaining

- **i.** Grow the cells until 80% confluence in the 4-well plates (Nunc).
- **ii.** Discard the growth medium. Wash the cells with 1 × PBS.
- **iii.** Fixation: Add 4% paraformaldehyde (PFA) solution to the cell and incubate at room temperature (RT) for 30 min. Then, remove the PFA solution and wash the cells with 1 × PBS twice.

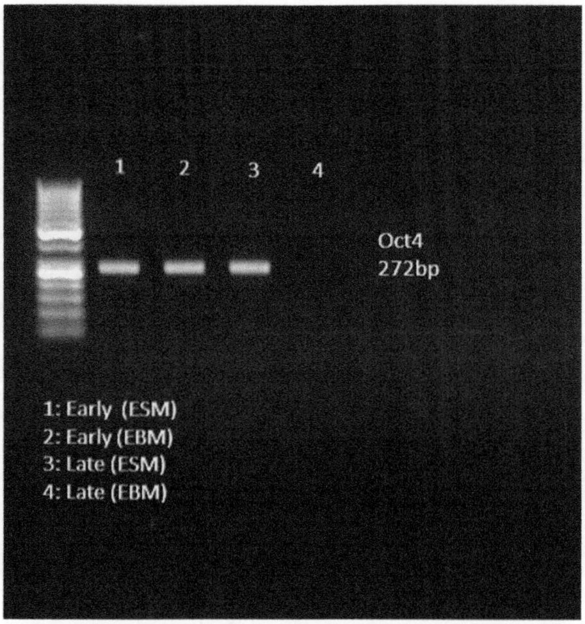

FIG. 9

Expression of Oct4 by RT-PCR. Oct4 expression is highly expressed in c-kit^{+} cells cultured at an early passage in the medium with (ESM) and without (EBM) serum and late passages in the medium with (ESM). There is no expression of Oct4 in cells upon prolonged culture in the absence of LIF (EBM) indicating the requirement of the leukemia inhibitory factor (LIF) for maintenance of the cells. The Oct4 (272bp) amplicon was determined using the 50bp DNA ladder.

iv. Permeabilization: Add the permeabilization solution (1% Triton-X in PBS) to the cells and incubate at RT for 15min. Then, remove the PFA solution and wash the cells with 1 × PBS twice.

v. Blocking: Add the blocking solution (0.3% BSA) to the cells and incubate the cells at RT for 30min before incubation.

vi. Indirect immunofluorescence: Add appropriate primary antibodies to the cells and incubate at 4.0°C, overnight. Then, rinse the cells with PBS and incubate them with appropriate secondary antibodies (such as Alexa Fluor 488 goat antirabbit IgG, goat antimouse IgG2b or goat antimouse IgG1). Incubate at RT in the dark for 2h.

vii. Add 1.0μg/μL propidium iodide (PI) or 5.0μg/mL DAPI to counterstain the nuclei for 10min and wash the cells three times with 1 × PBS.

viii. Keep the cells at 4.0°C in the dark until they are ready to be viewed under an inverted fluorescence microscope (Fig. 10).

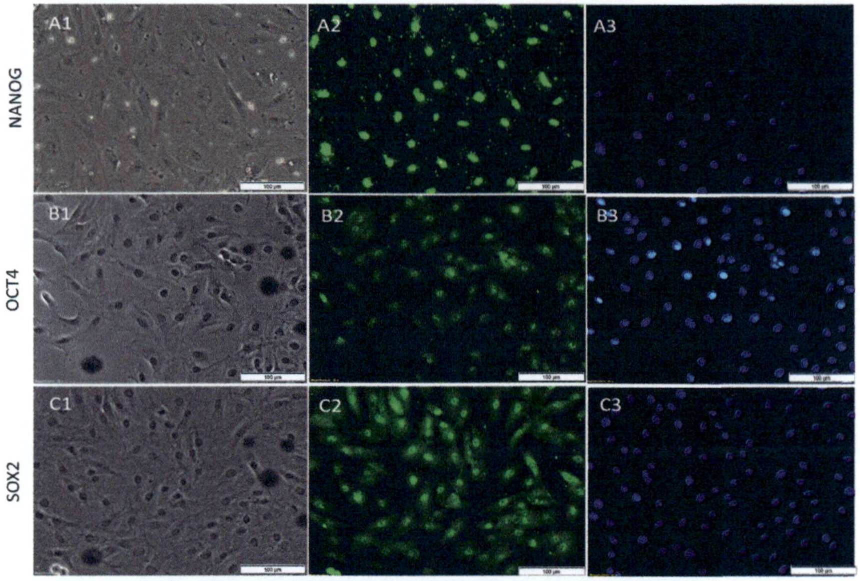

FIG. 10

Expression of intracellular markers by immunostaining. The morphology of R3 under phase contrast (A1, A2, A3) and after they were counter-stained with DAPI (A3, B3, C3), indicated by the blue fluorescence. The R3 clearly expressed pluripotency associated markers; Nanog, OCT4, and Sox2 (A2, B2, C3). The magnification is 20×.

(c) Flow cytometry

Intracellular markers (Oct4, Nanog and Sox2) (Fig. 11).

i. Grow the c-kit$^+$ cells until 80% confluence in a tissue culture flask.

ii. Remove the growth medium and wash the cell with 1× PBS.

iii. Detach the cells from the flasks by mild trypsinization. Centrifuge the cells at 160×*g* for 5min and discard the supernatant.

iv. Fixation and permeabilization step: Resuspend the cell pellets in ice-cold methanol. Incubate the cells for 40min at 4.0°C. Spin down the cells at 447×*g*, 4.0°C, and wash the cells with 1× PBS twice.

v. Indirect immunofluorescence: Incubate the cells with primary antibodies in 3% BSA for 1h at 4.0°C. Spin down the cells at 160×g, 4.0°C, and wash the cells with 1× PBS twice.

vi. Then, incubate the cells with the appropriate secondary antibody conjugated with IgG-APC, PE, or FITC for 30min in RT.

vii. Resuspend the cell pellets with PBS prior to flow cytometry analysis.

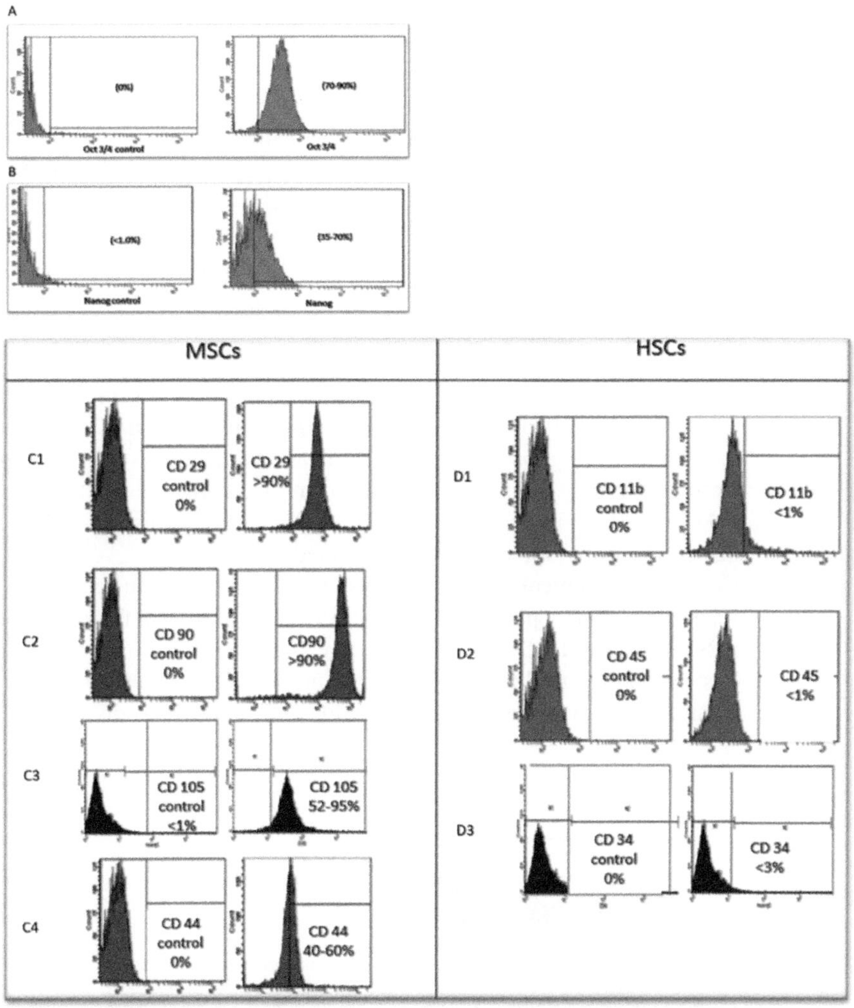

FIG. 11

Expression of intracellular and extracellular markers. FACS analysis reveals the Oct4 protein expression to be 70%–90% (A) and 35%–70% for Nanog (B). The associated mesenchymal markers CD29, CD90-1, CD105, and CD44 (C) were highly expressed. However, CD11b, CD45 and CD34 (D), the hematopoietic and endothelial markers, were not expressed in the c-kit$^+$ cells.

Extracellular markers: mesenchymal and adhesion molecule markers such as CD90-1, CD44, CD29, and CD71 and hematopoietic and endothelial markers such as CD45, CD11b (Fig. 11).

i. Repeat Steps i–iv.
ii. Direct immunofluorescence: Incubate the cells with antibodies conjugated with FITC/PE in 3% BSA for 1 h at 4.0°C. Spin down the cells at 160 × g, 4.0°C, and wash the cells with 1 × PBS twice.
iii. Resuspend the cell pellets with PBS prior to flow cytometry analysis.

5. *Spontaneous differentiation via the formation of multicellular aggregates, embryoid bodies (EBs).*
   (a) Spontaneous differentiation by the single-cell suspension method (Fig. 12)
      i. Grow the c-kit$^+$ cells 80% confluence in a tissue culture flask.
      ii. Remove the growth medium and wash the cell with 1 × PBS.
      Detach the cells from the flasks by mild trypsinization. Centrifuge the cells at 160 × g for 5 min and discard the supernatant. Count the cells.
      iii. Resuspend the cells in the EB medium (in the absence of LIF).
      iv. Add 2.5 × 10^6 cells in a low adherence 60.0 mm Petri dish (sterile bacteriological grade dish) in 5.0 mL medium.
      v. Change the EB medium every 2 days. This is done by transferring EB suspension using a plastic Pasteur pipette into a canonical flask. Let the EBs settle down for 5 min and remove the old EB medium using a plastic Pasteur pipette. Add the fresh EB medium into the EB suspension. Transfer the EB suspension in a new low adherence Petri dish.

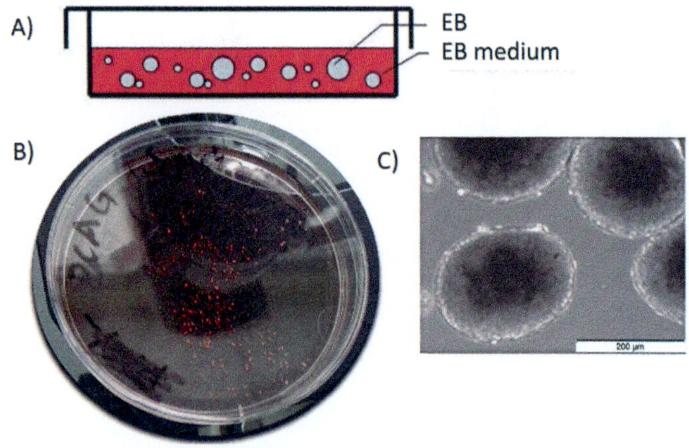

FIG. 12

EBs grown in cell suspension method. (A) A diagram and (B) a Petri dish plate showing the culturing of EBs in suspension medium viewed with naked eye. (C) The morphology of EBs formed via the suspension method viewed under 20 × magnification.

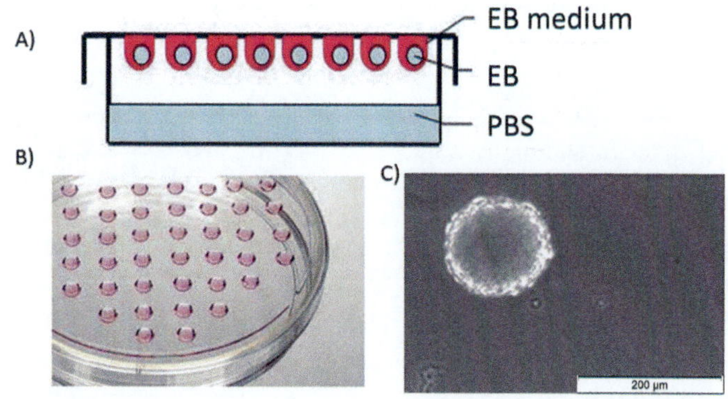

FIG. 13

EBs grown in hanging drop method. (A) A diagram and (B) a Petri dish plate showing the culturing of EB in drops of the medium hanging from the lid of the dish containing 1 × PBS. (C) The morphology of EB formed via the hanging drop method viewed under 10 × magnification.

- **vi.** On day 4, transfer the EBs onto a 4-well plate precoated with 0.1% gelatin, using a Pasteur pipette.
  View EBs under a phase-contrast inverted microscope.

(b) Spontaneous differentiation by the two steps hanging drop method (Fig. 13).
- **i.** Repeat Steps i–iv.
- **ii.** Using 100.0 μL, drop 20.0 μL of cell suspension, each containing 4000 cells, on the underside of the lid of a Petri dish. Incubate the culture in the incubator.
- **iii.** On day 2, collect and transfer the EB droplets to a Petri dish with the EB medium, pooling all the EB droplets. Let the EBs settle down for 5 min and remove the old EB medium using a Pasteur pipette. Add fresh the EB medium. Repeat Step iii.
- **iv.** On day 4, collect and transfer the EB droplets onto a 48-well plate precoated with 0.1% gelatin, using Pasteur pipettes.
- **v.** View EBs under a phase-contrast inverted microscope.

(c) Observation on EBs
- **i.** Observe EBs under different magnification to assess their quality.
- **ii.** Observe the size of EBs. The EBs look small from day 1 to 4 but mature after day 4. They should be about 100–300 μm when they are matured (Fig. 13).
- **iii.** Assess the boundary and cavitation process of the EBs. Mature EB exhibits smooth boundary and should be undergoing the cavitation process (Fig. 14).

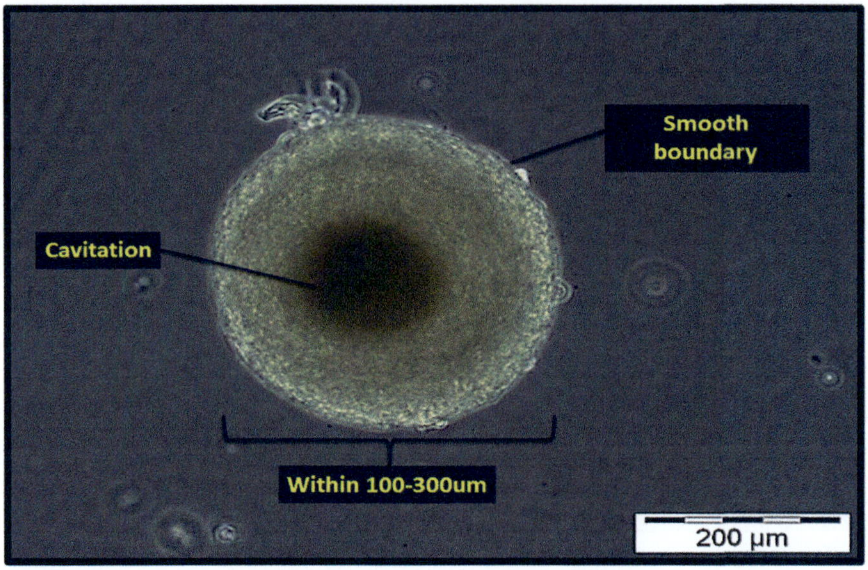

FIG. 14

A good quality EB. An EB should have a smooth boundary, undergo a cavitation process, and have a diameter within the range of 100–300 μm. A representative of EBs formed from c-kit$^+$ cells using the hanging drop and grown in a low adherence petri dish.

6. *Spontaneous differentiation of EBs into derivatives of the three primary germ layers.*
   a. Pull 10–20 EBs in a gelatin-coated well of a 4-well plate. Let the EBs differentiate and occupy the well for a few days.
   b. For ectodermal derivates such as neurons, stain the EB-differentiated cells with the neuronal marker (such as Tuj1) after 5 days in culture following the immunocytochemistry protocol (Fig. 15).
   c. Stain the EB-differentiated cells with the appropriate mesodermal marker (such as Brachyury) for mesodermal derivatives after 5 days in culture (Fig. 16).
   d. Stain the EB-differentiated cells with the appropriate endodermal marker (such as GATA4) for endoderm derivatives after at least 18 days in culture (Fig. 17).

3. Conclusion

Samples of the rat AF at full-term gestation can be procured from the 20th day of pregnancy (full-term) time-mated Sprague-Dawley rats. The primary culture is propagated before achieving an adequate number of cells for isolation.

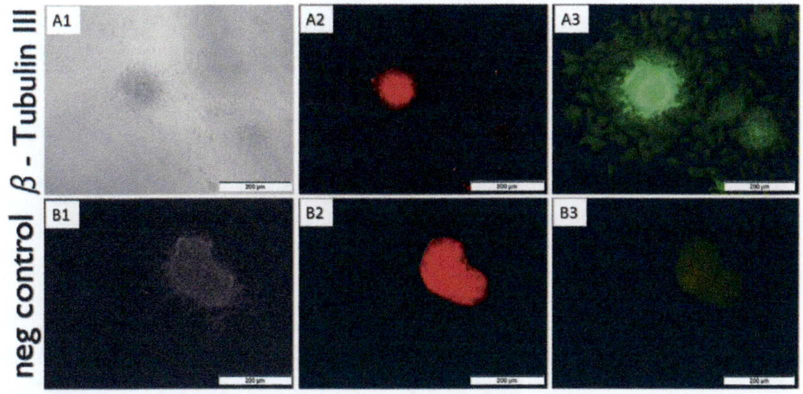

FIG. 15

Expression of ectoderm-specific marker, class III ß-Tubulin by 5th day EBs. The red fluorescence indicated the 5th-day EBs (A1, B1) were counterstained with propidium iodide (PI; A2, B2). The EBs clearly expressed β-tubulin III (A3). Staining the cells with only secondary antibodies showed a negative expression of the ectoderm markers (B3). The magnification is 10×.

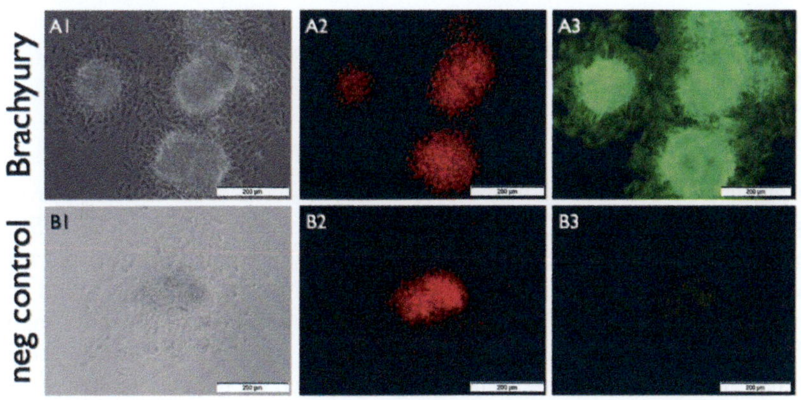

FIG. 16

Expression of the mesoderm-specific marker by 5th day EBs. The red fluorescence indicated the 5th-day EBs (A1, B1) were counterstained with propidium iodide (PI; A2, B2). The EBs clearly expressed mesoderm-specific marker, Brachyury (A3). Staining the cells with a secondary antibody only showed a negative expression of the mesoderm markers (B3). The magnification is 10×.

Immunoselection of c-kit$^+$ cells protocol is applied to isolate the c-kit+ amniotic fluid stem cells (AFSCs) (Mun-Fun et al., 2015). The molecular characterization for rat AFSCs at full-term gestation can be performed for pluripotent-associated markers and cell-surface markers associated with various stem cell types. Mun-Fun et al. (2015) found that a vital transcription factor for maintaining

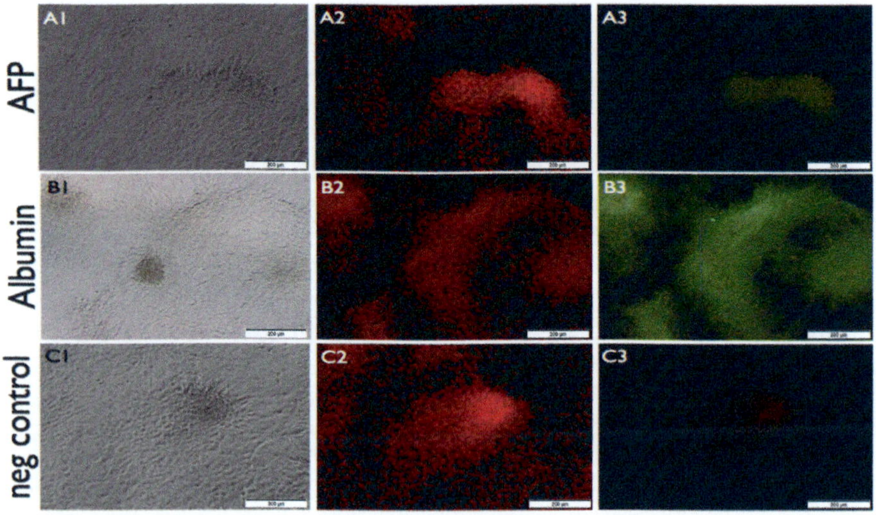

FIG. 17

Expression of the endoderm-specific marker by 18th day EBs. On the 18th day, EBs (A1, B1, C1) were counterstained with propidium iodide (PI; A2, B2, C2), indicated by the red fluorescence. Although the EBs showed no expression of AFP (A3), they clearly expressed albumin (B3). Staining the cells with the secondary antibody alone showed a negative expression of the endoderm markers (C3). The magnification is 10×.

pluripotent stem cells, Oct 4, was expressed in 70%–90% of the cells, as proved by immunocytochemistry (ICC) and fluorescence-activated cell sorting (FACS) analyses. Besides, its presence was also confirmed by RT-PCR. In addition, a vital transcription factor that regulates pluripotency, Nanog, was also found to be expressed by 35%–70% of the cells. Moreover, the expression of other markers such as SSEA-1 (embryonic stem cell surface marker), Sox2 (pluripotency marker), and Tert (the stemness-related gene that codes for telomerase) was also detected. Apart from this, other stem cells associated with cell-surface markers were also detected via FACS analysis. The markers include general stem cell markers CD90-1 and CD29 where 90%–99% and 80%–99% of the cells express them respectively, mesenchymal stem markers CD105 expressed highly by 51%–95% of the cells, and CD44 expressed moderately by 40%–60% of the cells. Nevertheless, endothelial and hematopoietic lineage-associated markers (CD34, CD11b and CD45) were not detected (Mun-Fun et al., 2015).

The functional characterization should also be performed to investigate the differentiation capability of c-kit$^+$ AFSCs in vitro and eventually prove the presence of highly potent stem cells among c-kit$^+$ cells. These criteria have been assessed by examining their ability to form embryoid bodies (EBs), spontaneously differentiate into derivatives of the three primary germ layers and directedly differentiate into neural lineage (Mun-Fun et al., 2015). EBs can be created using hanging drop and single-cell suspension methods and examined based on boundary smoothness, area size, and diameter. Both methods yielded good quality EBs with smooth boundaries and experienced cavitation process. These results parallel with standard pluripotent stem cells like mouse embryonic stem cells. Apart from this, lineage-specific markers for derivatives of the three primary germ layers are assessed to evaluate the ability of c-kit$^+$ AFSCs to differentiate spontaneously. These markers include Class III beta-tubulin (ectoderm), Brachyury (mesoderm), and albumin (endoderm). Class III beta-tubulin and Brachyury have been expressed after 5 days and albumin after 20 days in culture (Mun-Fun et al., 2015).

The established rat full-term amniotic fluid stem cells have been cultured up to 100 passages. The cells maintain their characteristics and differentiation potential as long as they are cultured or maintained appropriately in the presence of the rat leukemia inhibitory factor (LIF). The cells have been differentiated into neural lineages, cardiomyocytes, adipocytes, chondrocytes, osteocytes, and hepatocytes. It is worth noting that, in our hands, only rats sacrificed through cervical dislocation contained amniotic fluid cells that survived and propagated when cultured in vitro.

Their ability to differentiate into various cells has made them valuable for fundamental research. Their differentiated cells are also useful as the in vitro model for certain diseases, such as their differentiated neurons could be used as a model for neurodegenerative diseases.

References

Cananzi, M., Atala, A., Coppi, P.D., 2011. Stem cells from amniotic fluid. In: Principles of Regenerative Medicine. Elsevier, https:/doi.org/10.1016/B978-0-12-381422-7.10012-4.

De Coppi, P., Bartsch, G., Siddiqui, M.M., Xu, T., Santos, C.C., Perin, L., Mostoslavsky, G., Serre, A.C., Snyder, E.Y., Yoo, J.J., Furth, M.E., Soker, S., Atala, A., 2007. Isolation of amniotic stem cell lines with potential for therapy. Nat. Biotechnol. 25 (1), 100–106. https:/doi.org/10.1038/nbt1274.

Hamid, A.A., Joharry, M.K., Mun-Fun, H., Hamzah, S.N., Rejali, Z., Yazid, M.N., Karuppiah, T., Nordin, N., 2017. Highly potent stem cells from full-term amniotic fluid: a realistic perspective. Reprod. Biol. 17 (1), 9–18. https:/doi.org/10.1016/j.repbio.2017.02.001.

Miettinen, M., Lasota, J., 2005. R EVIEW neoplastic tissues and mutations and their. Rev. Lit. Arts Am. 13 (3), 205–220.

Mun-Fun, H., Ferdaos, N., Hamzah, S.N., Ridzuan, N., Hisham, N.A., Abdullah, S., Ramasamy, R., Cheah, P.S., Thilakavathy, K., Yazid, M.N., Nordin, N., 2015. Rat full-term amniotic fluid harbours highly potent stem cells. Res. Vet. Sci. 102, 89–99. https:/doi.org/10.1016/j.rvsc.2015.07.010.

Ramasamy, T.S., Velaithan, V., Yeow, Y., Sarkar, F.H., 2018. Stem cells derived from amniotic fluid: a potential pluripotent-like cell source for cellular therapy? Curr. Stem Cell Res. Ther. 13 (4), 252–264. https:/doi.org/10.2174/1574888X13666180115093800.

Further reading

Dobreva, M.P., Pereira, P.N.G., Deprest, J., Zwijsen, A., 2010. On the origin of amniotic stem cells: of mice and men. Int. J. Dev. Biol. 54 (5), 761–777. https://doi.org/10.1387/ijdb.092935md.

Hilmi, A.B.M., Fazliah, S.N., Fadilah, A.S., Asma, H., Razila, A.R.S., 2008. Stem cells from children's teeth. Arch. Orofac. Sci. 3 (1), 29–31.

Makino, S., Fukuda, K., 2011. Regenerating the heart. In: Cohen, I.S., Gaudette, G.R. (Eds.), Stem Cell Biology and Regenerative Medicine. Humana Press, https://doi.org/10.1007/978-1-61779-021-8.

CHAPTER

6 Human induced pluripotent stem cells (hiPSCs): Generation, characterization, and differentiation

Akon Higuchi[a,b,c]

[a]*State Key Laboratory of Ophthalmology, Optometry and Visual Science, Eye Hospital, Wenzhou Medical University, Wenzhou, Zhejiang, China,* [b]*Department of Chemical and Materials Engineering, National Central University, Taoyuan, Taiwan,* [c]*Center for Emergent Matter Science, Riken, Wako, Saitama, Japan*

1. Introduction

Human induced pluripotent stem cells (hiPSCs) and human embryonic stem cells (hESCs), which are called as human pluripotent stem cells (hPSCs), are valuable cell sources in cell therapies such as clinical treatment for many diseases. This is because hPSCs hold the specific ability to induce differentiation into all types of tissue cells in our human bodies (Higuchi et al., 2011, 2017, 2019). hESCs and hiPSCs that were induced to differentiation into beta cells secreting insulin and neurons secreting dopamine are promising to be used for the transplantation of these cells for the cure of diabetes and Parkinson's disease, respectively. The potential stem cell-based therapies and possible drug discovery platforms will be developed from pluripotent characteristics of hPSCs in future. Although hESCs are promising donor sources for cell transplantation therapies, they face immune rejection after their transplantation, and furthermore, hESC usage in the clinics holds ethical issues originated from the usage of human embryos. These drawbacks could be solved by using hiPSCs obtained directly from patients' tissue cells or shelve-off hiPSCs (Takahashi et al., 2007; Vodyanik et al., 2008). hiPSCs can be generated from tissue cells by the "forced" expression of pluripotency genes or proteins, such as Klf-4, Oct3/4, Sox2, and c-Myc (or L-Myc) or Oct3/4, Sox2, Nanog, and Ln28 (Takahashi et al., 2007; Vodyanik et al., 2008). hiPSCs' characteristics are considered to be similar to hESCs in many aspects, such as differentiation ability into many types of tissue cells, teratoma formation potential, embryoid body formation potential, chromatin methylation patterns, doubling time, and the expression of stem cell

Stem Cell Laboratory Techniques. https://doi.org/10.1016/B978-0-12-823729-8.00005-1

proteins or genes. In this chapter, we discuss recent development of hiPSC generation, characterization, and differentiation into some specific lineage of the cells.

2. Literature review

2.1 HiPSC generation method

Several novel methods have recently developed to enhance the likelihood of obtaining clinically useful hiPSCs. Several researchers have developed a genetic transfection method that uses nonintegrated vectors and nonviral vectors, as well as pluripotent proteins and miRNA (Higuchi et al., 2014c). For example, nonintegrating adenoviruses was utilized to transfer four reprogramming genes (0.1%–0.2% of iPSC generation efficiency). Two plasmids including the four reprogramming factors were transiently transfected into rodent cells (0.0001%–0.003% of iPSC generation efficiency) (Higuchi et al., 2014c). Soldner and his colleagues reported reprogramming of patient-derived fibroblasts utilizing Cre recombinase excisable viruses to deliver three factors (Oct3/4, Sox2, and Klf-4) (0.005% of hiPSC generation efficiency) (Soldner et al., 2009). Furthermore, Kaji and his colleagues developed viral-free and factor-removable hiPSCs utilizing a single plasmid including four factors that were bound with piggyBac transposons where the transgenes became active when the reprogramming was started (2.5% efficiency) (Kaji et al., 2009). Unfortunately, residual vector sequences could not be removed in this method, which would generate insertional mutations. Transgene- and vector-free hiPSCs could be successfully produced using oriP/Epstein-Barr nuclear antigen-1-based vectors (nonintegrating vectors) or piggyBac transposition, although the efficiency of hiPSC success is extremely low such as approximately 0.0005% (Woltjen et al., 2009), which may remove one obstruct to the clinical application of hiPSCs. The extremely low efficiency of the hiPSC generation is one of the barriers to utilize hiPSCs for the clinical treatment of the patients.

Recently, some distinct studies have suggested that reprogramming efficiency can be enhanced and that reprogramming kinetics can be accelerated by manipulating epigenetic status using small bioactive molecules (Fig. 1) (Higuchi et al., 2014c). However, most of studies have focused to generate murine iPSCs and not yet have studied direct reprogramming into hiPSCs from tissue cells utilizing solely epigenome-modifying small bioactive molecules.

2.2 Generation of HiPSCs using small bioactive molecules

Small bioactive molecules, which can target specific functions and signaling pathways, are useful molecules in the reprogramming processes of tissue cells into hiPSCs. The small bioactive molecules hold the possibility to play a key functions in the development of clinical applications toward translational medicine as well as the investigation of the basic biology of stem cells. These approaches would open the door to regeneration therapy using hiPSCs, which utilize functional and uniform

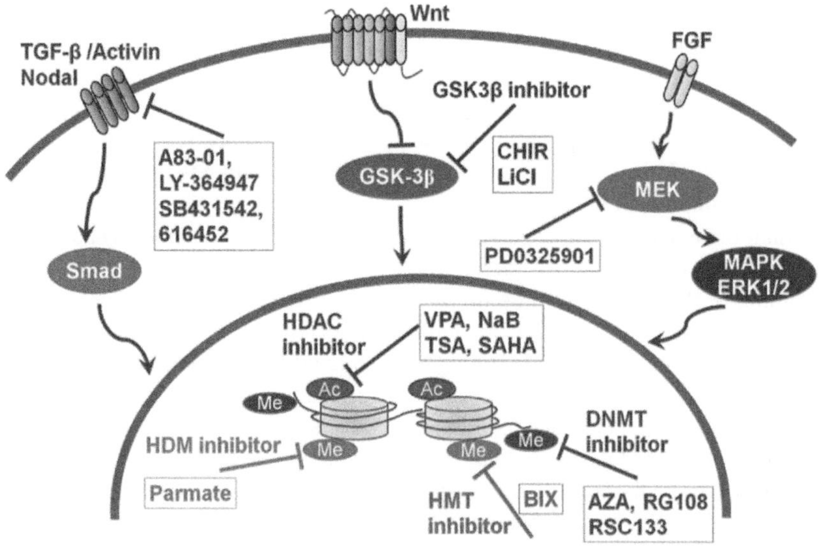

FIG. 1

The signaling pathways for stem cell fate and reprogramming using chemical manipulation. Small-molecule inhibitors of epigenetic modification enzymes, such as histone demethylases (HDM), histone methyltransferases (HMT), histone deacetylases (HADC), and DNA methyltransferases (DNMTs), act as small bioactive molecules for the reprogramming of tissue cells into hiPSCs (Higuchi et al., 2014c).

hiPSCs generated under chemically defined conditions (Higuchi et al., 2014c). Here, we discuss current situation developed in utilizing small bioactive molecules to maintain pluripotent states of the cells as well as to induce the reprogramming of tissue cells into hiPSCs.

We know that the expression of four transcription factors can induce reprogramming of some tissue cells into hiPSCs. Especially, hiPSCs are able to be reprogrammed by viral integration of Klf4/Sox2/Oct3/4 without expression of the oncogene c-Myc (currently L-Myc can be used instead of c-Myc), which has high possibility to generate tumor from hiPSCs. It is valuable to reduce the number of genes transduced into tissue cells to prepare hiPSCs in clinical usage. However, the decrease in the number of transcription factors leads to a decrease in iPSC success efficiency. Then, some small bioactive molecules were considered to contribute the enhancement of the efficiency of hiPSCs generation.

Epigenetic changes are necessary for reprogramming the cells into hiPSCs. Therefore, small molecular inhibitors toward epigenetic-changing enzymes play an important role in the epigenetic landscape. These enzymes are histone demethylases (HDMs), histone methyltransferases (HMTs), histone deacetylases (HDACs), and DNA methyltransferases (DNMTs), which work for the reprogramming of the tissue cells into hiPSCs (Higuchi et al., 2014c). The function of epigenetic modulators is the modification of chromatin structure and leads to be more permissive to transcriptional mechanisms during the reprogramming into hiPSCs. Several small bioactive molecules are listed up in Fig. 1 that target the transcription factors located on the downstream of the signaling pathways. The small bioactive molecules control the expression of transcription factors of pluripotency, guiding to the reprogramming of the cells into hiPSCs.

Small bioactive molecules having molecular weight less than 900 Da are able to freely diffuse through the cell membranes, which can be utilized to proceed reprogramming of the tissue cells into hiPSCs. Several researchers investigated systematically to discover small bioactive molecules to replace the transcription factors, which can work on the reprogramming or enhance the efficiency of the reprogramming. Typical high-throughput experiments utilized to identify small bioactive molecules are as follows: (1) Each well on the plates is inoculated with the cells, which are going to be transduced with Oct4 promoter that are driven enhanced GFP (green fluorescence protein), because the expression of endogenous Oct4 is the first marker where the cells are reprogrammed into hiPSCs. (2) Many small bioactive molecules are screened, which are able to replace Oct4 or Sox2. Small bioactive molecules, which can enhance the efficiency of reprogramming into hiPSCs, are also evaluated. The number of hiPSC colonies that express GFP is investigated as a read out of this process. The efficiency of colony formation is investigated from cells cultivated without and with small bioactive molecules.

Table 1 summarizes the small bioactive molecules, which are able to use the replacement of transcription factors, which are important to reprogram the cells into hiPSCs or to enhance the reprogramming rate of the cells into hiPSCs. These small bioactive molecules are blockers of signaling pathways on the reprogramming and epigenetic-modifying enzymes (Higuchi et al., 2014c).

2.3 Human iPSCs induced by small bioactive molecules in combination with a few transcription factors

The generation of mice iPSCs (miPSCs) is reported to be generated using small bioactive molecules by decreasing the numbers of transcription factors required (Higuchi et al., 2014c). Therefore, investigators considered to study the development of hiPSCs by decreasing the numbers of transcription factors with usage of bioactive small bioactive molecules. Some examples of the process of hiPSC preparation from human tissue cells are described in Fig. 2 (Higuchi et al., 2014c), and some of these studies for the reprogramming of human tissue cells into hiPSCs by transduction without and with transcription factors including small bioactive molecules are summarized in Table 2 (Giorgetti et al., 2009; Kim et al., 2009; Lee et al., 2012; Li et al., 2009; Takahashi et al., 2007; Wang et al., 2011; Zhu et al., 2010).

Just a few small bioactive molecules were found to enhance reprogramming efficiency and to replace transcription factors on hiPSC reprogramming compared

Table 1 Small bioactive molecules that facilitate reprogramming of tissue cells (Higuchi et al., 2014c).

Small bioactive molecules	Function as inhibitor or agonist	Transcription factors to be replaced or other function
RSC133, 5-azacytidine	DNA methyltransferase (DMNT) inhibitor	Reprogramming promotion
RG108	DNA methyltransferase (DMNT) inhibitor	Sox2 (with BIX) or Oct4, reprogramming promotion
Valproic acid (VPA), Trichostatin A (TSA), suberoylanilide hydroxamic acid (SAHA), sodium butyrate	Histone deacetylase (HDAC) inhibitor	Reprogramming promotion
LiCl	GSK-3β and LSD1 inhibitor	Reprogramming promotion
Compound B6	AKt-mediated GSK3-β inhibitor	Reprogramming promotion
Kenpaullone	GSK-3/CDKs inhibitor	Klf4
CHIR99021 (CHIR)	GSK-3β inhibitor that activate Wnt signaling pathway	Sox2
BIX-01294 (BIX)	G9a histone methyltransferase (G9a HMTase) inhibitor	Oct4
Tranylcypromine (Parnate)	H3K4 demethylation inhibitor	Reprogramming promotion
A83-01, SB431542	TGF-β inhibitor	Reprogramming promotion
LY-364947	TGF-β inhibitor	Sox2
616452 (E-616452, RepSox)	TGF-β inhibitor (ALK inhibitor II)	Sox2
Compound B4 (TGFb-RI)	ALK4 inhibitor	Reprogramming promotion
N-Oxaloylglycine	Prolyl-4-hydroxylase inhibitor	Reprogramming promotion
PD0325901 (PD)	MEK inhibitor	Reprogramming promotion
AMI-5	Protein arginine methyltransferase inhibitor	Sox2, Klf4 (with A-83-01)
PP1, iPY razine (iPY), Dasatinib	Src family tyrosine kinase inhibitor	Sox2
Bay K8644 (BayK)	L-channel calcium agonist	Sox2

Continued

Table 1 Small bioactive molecules that facilitate reprogramming of tissue cells (Higuchi et al., 2014c)—cont'd

Small bioactive molecules	Function as inhibitor or agonist	Transcription factors to be replaced or other function
Forskolin (FSK)	cAMP agonist	Oct4 with D4476 and 2-Me-5HT
D4476	Casein kinase 1 (CK1) inhibitor	Oct 4 with FSK and 2-Me-5HT
Compound B10	P38 kinase inhibitor	Reprogramming promotion
Compound B8	Inositol triphosphate 3-kinase (IP3K) inhibitor	Reprogramming promotion
Rapamycin	mTOR inhibitor	Reprogramming promotion
Rolipram, prostaglandin E2	cAMP agonist	Reprogramming promotion
TTNPB	Retinoic acid receptor ligand	Reprogramming promotion
2,4-Dinitrophenol (DNP)	Oxidative phosphorylation uncoupler	Reprogramming promotion
3-Deazaneplanocin (DZNep)	Epigenetic modulators	Reprogramming promotion
Quercetin	Hypoxia-inducible factor pathway activator	Reprogramming promotion
Fructose 2,6-bisphosphate	Phosphofructokinase 1 activator	Reprogramming promotion
8-Bromoadenosine 3″5″-cyclic monophosphate (8-Br-cAMP)	cAMP-dependent protein kinase activator	Reprogramming promotion
5-(4-Chloro-phenyl)-3-phenyl-pent-2-enoic acid (PS48)	PDK1 (3′-phosphoinositide-dependent kinase 1) activator	Reprogramming promotion
2-Methyl-5-hydroxytryptamine (2-Me-5HT)	5-HT3 agonist	Oct4 with D4476 and FSK
Shh, Purmorphamine, Oxysterol	Sonic hedgehog signaling	C-Myc, Klf4, and Sox2

to miPSC reprogramming. The molecules for hiPSC reprogramming are 2,4-dinitrophenol, quercetin, fructose 2,6-bisphosphate, 8-Br-cAMP, PS48 (PDK1 activator), *N*-oxaloylglycine, PD325901 (MEK inhibitor), SB431542 (TGF-β inhibitor), E-616452 (TGF-β inhibitor), LiCl, CHIR99021 (GSK3-β inhibitor), VPA, and sodium butyrate (HDAC inhibitor) (Table 1). It should be valuable to investigate

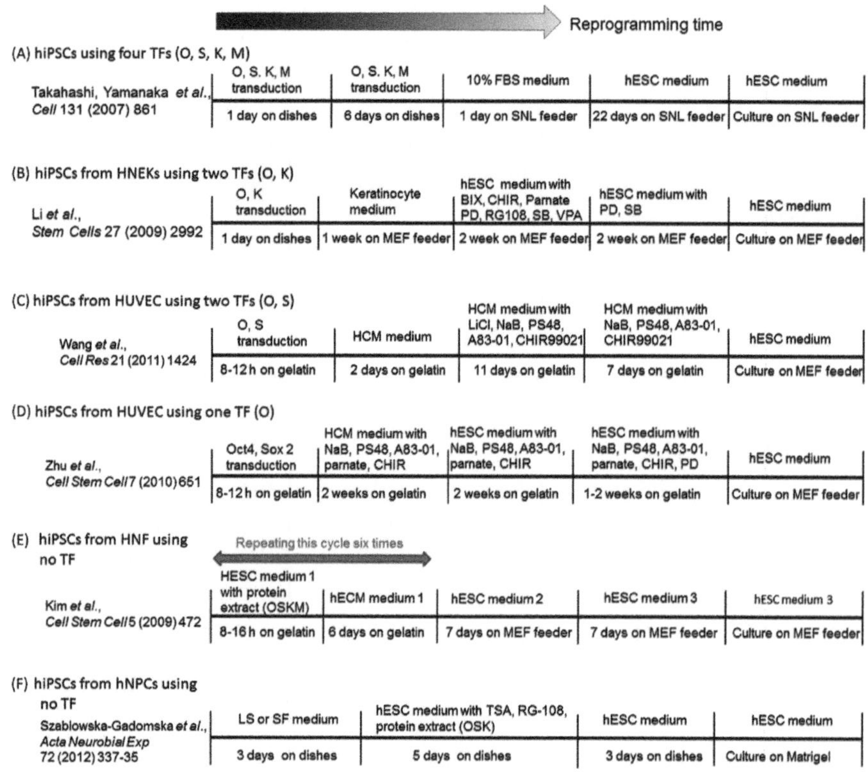

FIG. 2

Some examples of hiPSC preparation from human tissue cells by decreasing the number of transcription factors (TFs) with and without small bioactive molecules. M, K, S, and O indicate the transcription factors of c-Myc, Klf4, Sox2, and Oct4, respectively (Higuchi et al., 2014c).

whether other bioactive molecules having biochemical potential during the reprogramming of miPSCs can be used for the reprogramming of hiPSCs. In present, the Oct4 transduction in combination with small bioactive molecules is minimum required to make hiPSC generation. However, the success rate of hiPSC reprogramming is extensively low compared to that of miPSC reprogramming (approximately 0.0001%–0.02%).

Neural stem cells (NSCs) endogenously show expression of c-Myc, Klf4, and Sox2 to some extent. Therefore, Kim and his colleagues investigated the reprogramming of human fetal NSCs into hiPSCs by using transcription factor of only Oct4 (Kim et al., 2009). This study indicates that only one transcription factor of OCT4 seems to be enough for the reprogramming of the cells into hiPSCs. It should

Table 2 Reprogramming of human tissue cells into hiPSCs by transduction without and with transcription factors by the aid of small bioactive molecules (Higuchi et al., 2014c).

Transcription factors required[a]	Somatic cells[b]	Small bioactive molecules	Efficiency (%)	Ref. (year)
Four factors				
MKSO	HFFs	None	0.02%	Takahashi et al. (2007)
MKSO	HFFs	RSC133	Three times higher than control	Lee et al. (2012)
Two factors				
KO	HF-NPCs	None	0.01%	Kim et al. (2009)
SO	HCBSCs	None	0.01%	Giorgetti et al. (2009)
KO	HNK	Parnate, CHIR99021 (and SB431542, PD0325901)	0.005%–0.01%	Li et al. (2009)
KO	HNK	PS48, NaB, PD032590, A-83-01 (and CHIR99021, Parnate)	0.03%	Zhu et al. (2010)
SO	HUVECs	LiCl, CHIR88021, PS48, NaB, A83	0.00%	Wang et al. (2011)
One Factor				
O	HF-NPCs	None	0.00%	Kim et al. (2009)
O	HNK	PS48, sodium butyrate, PD032590, A-83-01 (and CHIR99021, Parnate)	0.00%	Zhu et al. (2010)
O	AFCs and HUVECs	PS48, NaB, PD032590, A83-01 (and CHIR99021, Parnate)	0.0025%–0.01%	Zhu et al. (2010)
O	HUVECs	LiCl, PS48, NaB, A83-01, CHIR99021 (or PD0325901)	0.00%	Wang et al. (2011)

[a]*M, K, S, and O indicate c-Myc, Klf-4, Sox2, and Oct4, respectively.*
[b]HF-NPCs, *human fetal neural progenitor cells;* HNK, *human neonatal keratinocytes;* AFCs, *amniotic-derived cells;* HUVECs, *human umbilical vein endothelial cells;* HFFs, *human foreskin fibroblasts;* HCBSCs, *human cord blood-derived stem cells.*

be noted that it is extremely hard to get human fetal NSCs to make patient-specific hiPSCs, and the success rate of hiPSC reprogramming is obtained to be extremely low (0.005%) in this investigation.

Zhu and his colleagues established their four-step processes to make hiPSCs from amniotic-derived cells (AFCs), human umbilical vein endothelial cells (HUVECs), and human neonatal keratinocytes (HNKs) (Zhu et al., 2010). In the generation of hiPSCs derived from HUVECs, the following method was utilized: (i) In the first step, the cells were transduced with Oct4 on ECM-coated plates for approximately 10h. (ii) In the second step, the cells were cultivated in HUVEC media including CHIR99021, tranylcypromine, A83-01, PS48, and sodium butyrate for 14days on the gelatin-coated plates. (iii) In the third step, the cells were cultivated in hPSC media including CHIR99021, tranylcypromine, A83-01, PS48, and sodium butyrate for another 14days on the gelatin-coated plates. (iv) In the fourth step, the cells were cultivated in hPSC media including PD0325901, CHIR99021, tranylcypromine, A83-01, PS48, and sodium butyrate for another 7–14days on the gelatin-coated plates, and the reprogrammed hiPSCs were compared with the conventional hiPSCs using four transcription factors (Fig. 2D). The single transcription factor, Oct4, with small bioactive molecules is able to reprogram some human tissue cells into hiPSCs, which are the same to pluripotent hESCs from the aspects of their colony formation and differentiation potential as well as protein and gene expression. However, the efficiency of the reprogramming into hiPSCs was reported to be extremely low (0.005%) in this case.

Wang and his colleagues prepared hiPSCs utilizing either one transcription factor (Oct4) or two transcription factors (Sox2 and Oct4) using small bioactive molecules (Wang et al., 2011). Their small bioactive molecules were similar to the molecular combination utilized by Zhu and his colleagues (Zhu et al., 2010). However, Wang and his colleagues used LiCl (5mM) from days 3 to 18, which could enhance the efficiency of miPSC and hiPSC reprogramming (Fig. 2C) (Wang et al., 2011). Human iPSC colonies were selected for proliferation on mouse embryonic fibroblasts (MEFs) after 30days of cultivation. The effect of LiCl on enhancing the efficiency of iPSC generation depended on its major target of GSK-3β partially. On the other hand, LiCl is not like other GSK-3β blockers, because LiCl can promote expression of Nanog and enhances its transcriptional potential. LiCl can serve as enhancing epigenetic modifications from the downregulation of LSD1, an H3K4-specific histone demethylase. However, the success rate of reprogramming into hiPSC was extensively low (0.0005% or 0.0015% for one-factor or two-factor transduction, respectively) in this study.

If hiPSCs could be made from patient-specific or allogenic human tissue cells utilizing only small bioactive molecules with reasonable efficiency (e.g., >0.05%), the doors will be open for the possible usage of hiPSCs in clinics. At present, reprogramming of the tissue cells into hiPSC, which are solely based on small bioactive molecules has not yet been succeeded, although hiPSCs seem to be prepared utilizing proteins or cell extracts with no usage of transduction of transcription factors.

2.4 Cell culture biomaterials for hiPSC culture

The most popular method to maintain hPSCs in a pluripotent state is to cultivate hiPSCs on feeder cells such as (a) MEFs, (b) human feeder cells, or (c) on Geltrex or Matrigel where Geltrex and Matrigel are extracts from the sarcomas of Engelbreth-Holm-Swarm mice that contain growth factors (e.g., TGF-β, FGF, and EGF), enactin, heparan sulfate proteoglycans, collagen IV, and laminin (LN) (Fig. 3) (Higuchi et al., 2014b).

The feeder cell usage to cultivate hPSCs is laborious and time-consuming issue. In contrast, Geltrex and Matrigel can maintain the pluripotency of most of hPSC lines, although the cultivation of hPSCs on Geltrex- and Matrigel-coating plates is undefined, and their xenogenic conditions, which are not favor for the clinical therapy using hPSCs. Especially, there is a possibility of transfer of xenogenic viruses and the generation of immunogenic epitopes to humans (e.g., nonhuman sialic acid *N*-glycolylneuraminic acid (Neu5Gc)).

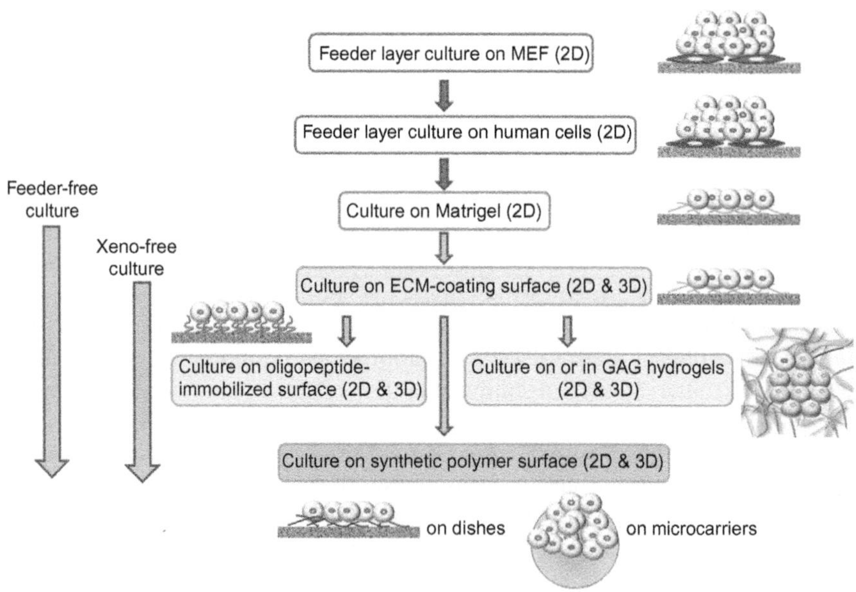

FIG. 3

Human PSC cultivation. Human PSC are able to be cultivated on MEFs or human feeder cells. Human PSCs are possible to culture on Matrigel in xeno-containing conditions. Human PSCs are able to be cultured on ECM-coated or oligopeptide-immobilized plates in xeno-free and feeder-free conditions. Human PSCs are also possible to be cultivated on polysaccharide gels and on synthetic polymetic plates, when optimal polysaccharide or polymer are selected (Higuchi et al., 2014b).

It is important to investigate mass production conditions of hPSCs under xeno-free and feeder-free conditions in the facilities that hold cGMP (current good manufacturing practice). The usage of feeder cells on hPSC cultivation restricts the usage of hPSCs. Several researchers have reported some alternative hPSC cultivation conditions, which are not necessary to use feeder cells. The key points being responsible to maintain hPSC proliferation are not currently known exactly (Higuchi et al., 2014b). The activin/nodal, Wnt, and basic FGF (FGF-2) signaling pathways should be considered to be critical to maintain the pluripotent states of hPSCs. The proliferation under maintaining hPSC pluripotency is a valuable engineering to create the bankable sources of hiPSCs or hESCs, which are necessary for future regenerative therapies.

Currently, some biomaterials are designed to cultivate hPSCs and to keep their pluripotency together with chemically defined cultivation media. These biomaterials are typically made of nonxenogenic materials under chemically defined conditions. However, the designed biomaterials need a combination of specific cultivation medium in general. Currently, typical cell culture biomaterials used are recombinant vitronectin (VN)-coated plates, LN521-coated plates, and Synthemax-coated plates, which are grafted with oligopeptide derived from VN.

Some researchers have reported the effects of cell cultivation biomaterials on stem cell fate of hPSCs. Especially, since 2010, some novel matrices and protocols for hiPSC and hESC cultivation on the biomaterials are reported under xeno-free and feeder-free cultivation conditions. Several strategies are considered to develop cell culture matrices for hPSC cultivation under xeno-free and chemically defined conditions. These strategies are hiPSC and hESC cultivation (a) in porous or hydrophilic three-dimensional (3-D) microcapsules, (b) on 3-D microcarriers without or with ECM immobilization, (c) on two-dimensional (2-D) dishes prepared from synthetic monomers, (d) on 2-D hydrogels prepared from polysaccharide like GAG (glycosaminoglycan), (e) on 2-D plates with immobilized synthetic oligopeptides obtained from ECMs, and (f) on 2-D plates with immobilized natural ECMs (Figs. 3 and 4) (Higuchi et al., 2014b).

2.5 HiPSC cultivation on ECM-immobilized biomaterials

It is reasonable to consider that 2-D cultivation of hiPSCs and hESCs is based on the natural ECMs secreted by the feeder cells for their adhesion and expansion. Therefore, investigators started to utilize the ECM components obtained from feeder cells. Two types of ECMs are typically utilized for hiPSC and hESC cultivation: (a) mixed components or single component from chemically defined ECMs (VN, LN, and fibronectin) and (b) decellularized ECMs prepared from human fibroblasts that are xeno-free but chemically undefined. The ECMs are typically utilized as coating molecules on cell cultivation scaffolds or plates. This is because Matrigel is possible to maintain the pluripotency of hiPSCs and hESCs without the usage of feeder cells where Matrigels contain ECMs such as heparan sulfate proteoglycans, collagen IV, and LN (Higuchi et al., 2014b).

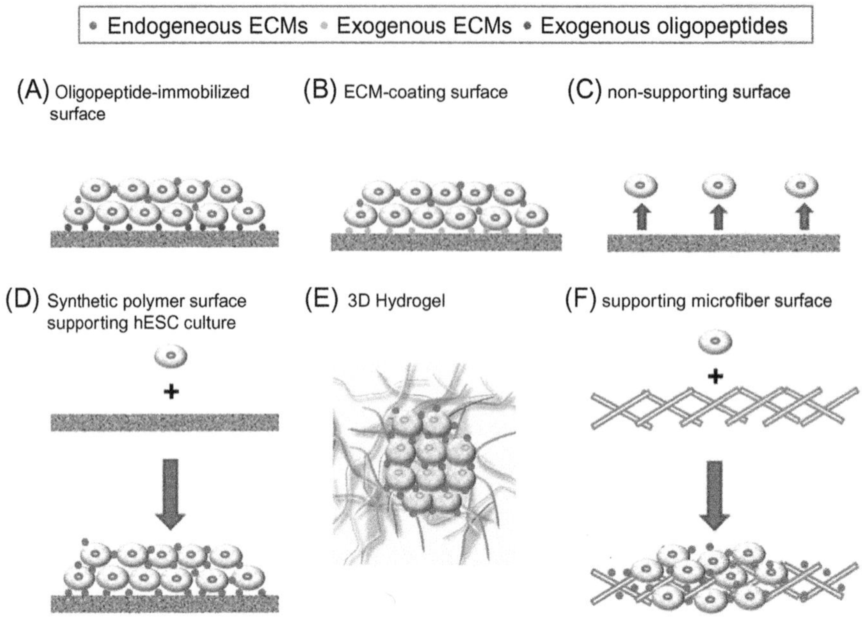

FIG. 4

Illustration of hPSC cultivation. hPSC cultivation on (A) oligopeptide-immobilized surfaces (2-D and 3-D cultivation), (B) ECM-immobilized surfaces (2-D and 3-D cultivation), (C) surfaces, which are not suitable for hPSC cultivation (2-D and 3-D cultivation), (D) synthetic polymeric surfaces, which can keep pluripotency of hPSCs (2-D and 3-D cultivation), (E) 3-D hydrogels prepared with GAG or another polymeric materials, and (F) microfiber surfaces, which are suitable for hPSC cultivation (Higuchi et al., 2014b).

Using a freeze-thaw procedure, Fu and his colleagues prepared ECMs obtained from decellularized feeder cells, which were taken from embryonic bodies that were differentiated from hESCs (Fu et al., 2011). ECMs that were composed of fibronectin, LN, and collagen IV were utilized for the cultivation of hESCs in xeno-free TeSR2 media for over 18 passages. Then, this research indicates a new idea for the efficient culture of clinically compliant hPSCs in xeno-free and feeder-free conditions.

Typical hPSC lines were originally prepared via cultivation on MEFs at the initial development. Then, the cell lines are not suitable to utilize for the clinical application to treat human patients. Therefore, Ilic and his colleagues established hPSCs on decellularized fibroblasts from human neonatal foreskin under xeno-free conditions (Ilic et al., 2012). Their specific mutation-carrying (Myotonic dystrophy 1 or Huntington's disease) hESC lines were established and cultivated on the decellularized ECMs. However, plates immobilized with CELLstart (composition of human serum albumin and fibronectin) could not maintain the pluripotency and proliferation of these hESCs in StemPro media including activin A, IGF1, heregulin, and FGF-2, although conventional hESCs can proliferate with their pluripotency on CELLstart in the StemPro media. Therefore, the combination of a specific cultivation media and cell cultivation biomaterials should be extensively chosen, which depend on the properties of specific hPSCs. Currently, it is difficult to find a perfect coating materials or cell cultivation plates, which can maintain any types of hPSC in a pluripotent state for long time (i.e., more than 20 passages), although recombinant VN-coated plates are typically used for hPSC cultivation.

Natural or recombinant VN or LN have been utilized instead of decellularized ECMs and Matrigels for the feeder-free cultivation of hiPSCs and hESCs because their chemical properties are well defined. The ECM components, which can be immobilized on cell culture plates for hiPSC and hESC expansion as well as their attachment sites for hPSCs, are shown in Table 3 (Higuchi et al., 2014b).

LN is the first ECM developed in 2- to 4-cell-stage embryos and is a main molecule of the ECMs of the basal laminae in vertebrates (Higuchi et al., 2011). LN is a family of heterotrimeric glycoproteins, which are constructed of γ, β, and α chains, and form 15 combinations in human body. LN-521 is composed of the γ1, β2, and α5, segments and is typically used as a coating molecule on the plates utilized to cultivate hPSCs.

Several hPSC lines are expressing LN-511 chains, whereas hPSCs do not express LN-332. Miyazki and his colleagues reported that hPSCs expressed integrin α6β1 (Miyazaki et al., 2008), which binds extensively to LN-511/-521, LN-332, and LN-111 (Fig. 5) (Higuchi et al., 2014b). hPSCs were found to be cultivated on dishes immobilized with three recombinant LN in condition medium of MEF (MEF-CM) for several passages while keeping their pluripotent states. These findings indicated that recombinant LN-511, LN-332, and LN-111 can support the proliferation of

Table 3 Interaction between ECMs and integrin binding sites for attachment, proliferation, and differentiation of stem cells.

ECM	Binding site of cells
Collagen I	Integrin (αVβ3, α3β1, α2β1, α1β1)
Collagen II	Integrin (α10β1, α2β1, α1β1)
Collagen IV	Integrin (CD44, α2β1)
Fibronectin	Integrin (αVβ5, αVβ6, αllbβ3, αVβ3, α5β1, α4β1)
Vitronectin	Integrin (αVβ3, αVβ5)
Laminin	Integrin (α6β4, α6β1, α3β1, α2β1, α1β1)
Laminin-1 (laminin 111)	Integrin (α9β1, α7β1, α6β1, α2β1, α1β1), heparan sulfate proteoglycan, α-dystroglycan, sulfate
Laminin-5 (laminin 332)	Integrin (α6β4, α6β1, α3β1, α2β1)
Laminin-511	Integrin (α6β1)
Laminin-10/11	Integrin (α6β4, α6β1, α3β1)

Copyright 2014 (Higuchi et al., 2014b). Adapted with permission from Elsevier Ltd.

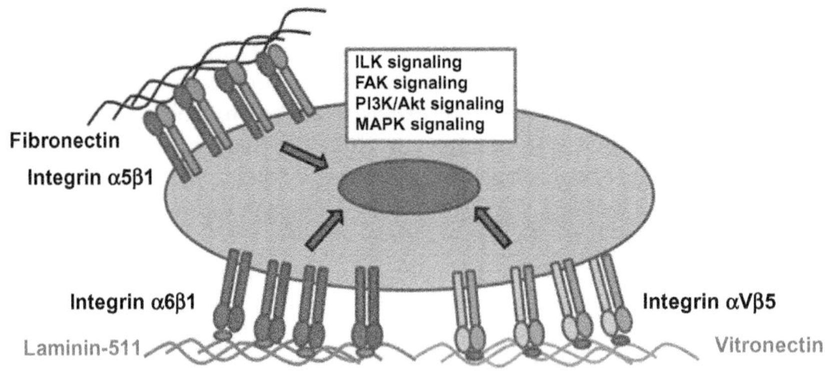

FIG. 5

Human PSC attachment on an ECM-coated surface. The attachment of hPSCs to laminin-511, vitronectin, and fibronectin is mediated by α6β1, αVβ5, and α5β1, respectively (Higuchi et al., 2014b).

pluripotent hPSCs because of their high affinity to integrin α6β1 that is found on hPSCs. However, this study was examined utilizing xeno-containing cultivation medium, and the hPSCs were only cultivated for a few passages. It is therefore difficult to find whether the hPSC pluripotency is able to be kept for more passages in xeno-free cultivation media. However, Rodin and his colleagues investigated that hESCs (HS401, HS207, and HS420) cultivated on LN-511-immobilized plates could keep their pluripotency for more than 18 passages holding a normal karyotype in either H3 media or O3 media (xeno-free media) (Rodin et al., 2010). In our laboratory, we typically use LN-521-coated dishes for hPSC culture and differentiation into cardiomyocytes. Especially, LN-521-coated dishes are appropriate surface for differentiation of hPSCs into cardiomyocytes. Furthermore, strong adhesion of hPSCs can be found on LN-511-coated dishes compared with recombinant LN-521-coated or VN-coated dishes, which hinders the passage process of hPSCs from our experience.

VN is known to be a glycoprotein (75 kDa) that is consisted of 460 amino acid sequences. Braam and his colleagues studied hPSC adhesion on plates immobilized with some ECMs (collagen IV, fibronectin, VN, and LN-111) at concentrations of 10–50 μg/mL (Braam et al., 2008). Investigation by Integrin-blocking indicated that (a) α6β1 integrin is considered to mediate the binding to LN-511, LN-332, and LN-111, (b) α5β1 integrin is considered to mediate the binding to fibronectin, and (c) αVβ5 integrin seems to mediate the binding of hPSCs to VN (Fig. 5) (Higuchi et al., 2014b). The recombinant VN-immobilized plates supported the proliferation of undifferentiated hPSCs for more than eight passages in mTeSR1 media in this investigation. Some other investigators also found that recombinant VN-immobilized plates support undifferentiated hPSC proliferation including our group. Yap and his colleagues investigated a threshold surface density of human plasma VN for the expansion of hPSCs (HES-3 and H1), which maintain their pluripotency

(Yap et al., 2011). The extensive expansion of hPSCs was reported for more than 20 passages on plates immobilized with >5 μg/mL VN in mTeSR1 media, which indicated 250 ng/cm^2 of VN as evaluated using the Bradford assay.

Meng and his colleagues studied the synergistic effects of the biomaterials, cell cultivation media, and exogenous factors on the growth and attachment of hiPSCs and hESCs under xeno-free and chemically defined cultivation media where hiPSCs and hESCs can keep their pluripotency for a long period (Meng et al., 2011). They suggested that hiPSC and hESC cultivation on VN-immobilized plates in TeSR2 media should be the optimal conditions to maintain long-term cultivation of two hiPSC lines and three hESC lines, whereas hiPSC and hESC cultivation on fibronectin-immobilized plates was found to be the second selection. From our experience, VN, in concert with a suitable cell cultivation media, is considered to be one of the optimal biomaterials to culture hPSC currently.

3. Characterization of hiPSCs

Human PSCs show expression of some surface markers of pluripotency, such as tumor rejection antigen 1-81 (Tra-1-81), Tra-1-60, Nanog, Oct-4 (POU5F1, POU domain transcription factor), Oct3/4, glycolipid stage-specific embryonic antigen 4 (SSEA-4), and SSEA-3 but not SSEA-1 (Higuchi et al., 2011, 2014b). Undifferentiated mouse ESCs and iPSCs express SSEA-1 but SSEA-1 is a differentiated marker for hESCs and hiPSCs. hPSCs also express high level of pluripotency genes, such as hTERT, Rex-1, Sox-2, Nanog, Oct-4, and Oct3/4. A characterization assay of pluripotent hESCs and hiPSCs is described in Fig. 6 (Higuchi et al., 2011, 2014b).

The pluripotency of hESCs and hiPSCs is analyzed from (a) the colony formation evaluated using phase inverted microscopy where hESCs and hiPSCs form colonies with tight boundaries; (b) the alkali phosphatase activity (ALP) evaluated using colorimetry detection kit; (c) the expression of pluripotency genes analyzed by qRT-PCR and RT-PCR experiments; (d) the expression of pluripotent proteins evaluated using immunohistochemical analysis and flow cytometry; and (e) the potential to induce differentiation into three germ layer (mesoderm, ectoderm, and endoderm) cells analyzed from teratoma formation in vivo and embryoid body (EB) formation in vitro, which can be evaluated from immunohistochemical assay or hematoxylin and eosin staining (H and E staining) (Fig. 6) (Higuchi et al., 2011, 2014b). It is further valuable to evaluate that the hESCs and hiPSCs show any abnormalities of karyotyping after long-term cultivation.

4. Materials, reagents, chemicals, apparatus and kits

The materials including reagents, chemicals, and kits are shown in Table 4.

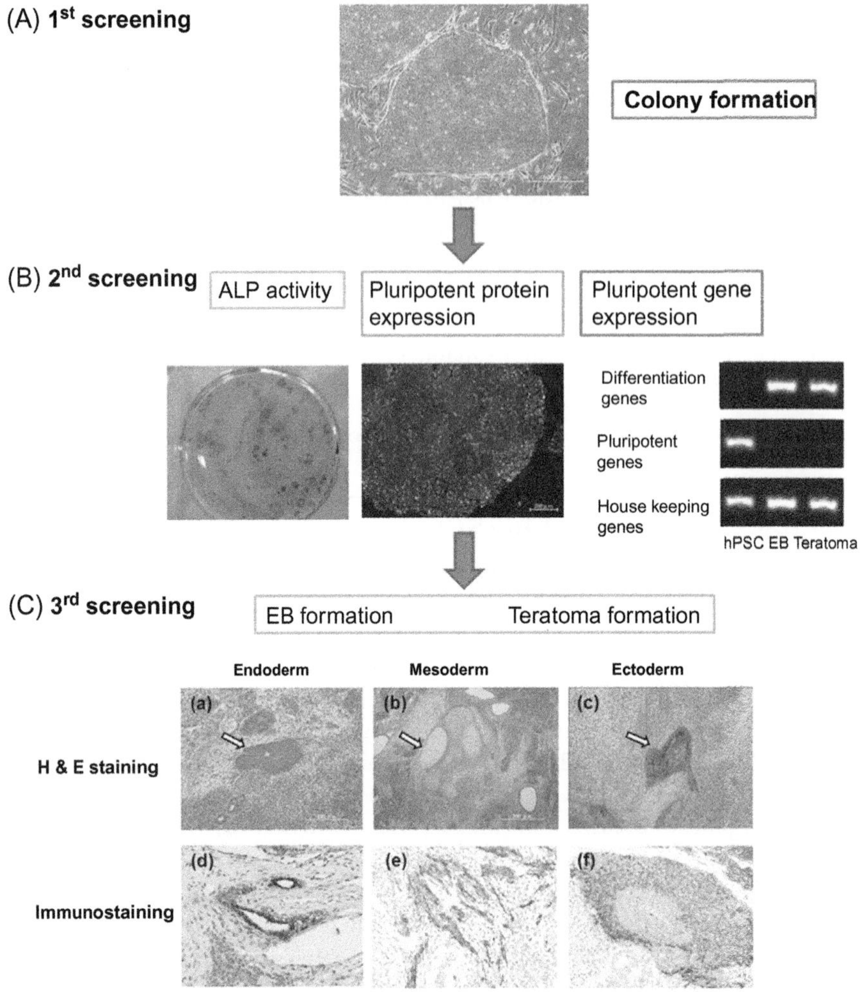

FIG. 6

Characteristic assay of hPSCs in pluripotent state. The pluripotency of hPSCs is analyzed from (A) the colony generation evaluated from phase inverted microscopy (first check point); (B) the ALP activity evaluated from an enzyme-substrate reaction, the pluripotent protein expression evaluated from immunostaining method or flow cytometry assay, and the pluripotent gene expression evaluated from qRT-PCR or RT-PCR assay (second check point); and (C) the potential to induce differentiation into cells from three germ layers evaluated from teratoma or EBs formation (Higuchi et al., 2011).

Table 4 Materials utilized in this research.

Materials	Abbreviation	Catalog No.	Company[a]
ECM			
Collagen type I	COL	354231	BD
Cellstart	Cellstart	A1014201	TFS
Fibronectin	FN	356008	BD
VN (recombinant)	rVN	A14700	TFS
LN-521	LN-521	LN521	BL
LN-511	LN-511	892021	NI
Matrigel	Matrigel	#356230	CO
Cell cultivation plates			
6-Well polystyrene plate	6-Well PS	353046	CO
Ultra-low attachment plate	Ultra-low attachment plate	3471	CO
Chemicals			
CHIR99021	CHIR99021	SML1046	SA
IWR-1	IWR-1	I0161-5MG	SA
A83-01	A83-01	SML0788	SA
BMP4	BMP4	SRP3016	SA
Alcian blue solution	Alcian blue	101647	MK
Dispase II	Dispase	D4693-1G	SA
Cell culture media and components			
Dulbecco's modified Eagle's media	DMEM	D5648	SA
Proliferation media	MCDB201	M6770-1L	SA
MesenCult-ACF media	ACF	#05440	SCT
StemPro MSC SFM media	StemPro XF	A1067501	TFS
DMEM/F12 media	DMEM/F12	11330-057	TFS
Essential 6 media	Essential 6	A1516401	TFS
Essential 8 media	Essential 8	A1517001	TFS
RPMI 1640 media	RPMI 1640	11875093	TFS
αMEM media	αMEM	M4526-500ML	SA
B27	B27	17504044	TFS
B27 minus insulin	B27 minus insulin	A1895601	TFS
Alkaline Phosphatase Assay Kit	SensoLyte pNPP	AS-72146	AS
Fetal bovine serum	FBS	04-001-1A	BI
Fibroblast growth factor basic	FGF-2	CYT-218	PST
Cell dissociation Kit	Cell Dissociation Kit	05426	SCT
Hoechst 33342	Hoechst	PA-3014	LO
Antibodies			
Alexa Fluor 594 donkey antirabbit IgG	Alexa Fluor 594 donkey antirabbit IgG	A-21207	LT

Continued

Table 4 Materials utilized in this research—cont'd

Materials	Abbreviation	Catalog No.	Company[a]
Alexa Fluor 594 goat antimouse IgG	Alexa Fluor 555 goat antimouse IgG	A-21424	TFS
Alexa Fluor 488 goat antirabbit IgG	Alexa Fluor 488 goat antirabbit IgG	A-11008	TFS
Mouse IgG1 Isotype	Isotype-control	MA5-14453	TFS
Anti-cTnT IgG	Anti-cTnT IgG	MA5-12960	TFS
Anti-MLC2V IgG	Anti-MLC2V IgG	ab92721	AB
Anti-α-actinin IgG	Anti-α-actinin IgG	A7811	SA
Anti-GFAP IgG	Anti-GFAP IgG	MA5-15086	TFS
Anti-SMA IgG	Anti-SMA IgG	PA5-19465	TFS
Anti-AFP IgG	Anti-AFP IgG	PA5-21004	TFS
Anti-Nanog IgG	Anti-Nanog antibody	23D2-3C6	TFS
Anti-SSEA-4 IgG	Anti-SSEA-4 IgG	ab16287	AB
Alexa Fluor 488 goat antimouse IgG	Alexa Fluor 488 goat antimouse IgG	A-11001	TFS
Anti-Sox2 IgG	Anti-Sox2 IgG	AB5603	MK
Anti-Oct3/4 IgG	Anti-Oct3/4 IgG	sc-5279	SCB
7-AAD Viability dye	7-AAD	559925	BD
PE Mouse IgG1κ, isotype control	PE isotype	555749	BD
FITC Mouse IgG1κ, isotype control	FITC isotype	555748	BD
FITC Mouse IgG2bκ, isotype control	FITC isotype	555742	BD
FITC mouse antihuman CD34	FITC anti-CD34	555821	BD
FITC mouse antihuman CD44	FITC anti-CD44	555478	BD
PE mouse antihuman CD73	PE anti-CD73	550257	BD
PE mouse antihuman CD90	PE anti-CD90	555596	BD
Reprogramming and differentiation kit			
CytoTune-iPS 2.0 Sendai Reprogramming Kit	CytoTune	A16518	TFS
Mice			
NOD.CB17-Prkdcscid/Jnarl	NOD-SCID	NOD.CB17-Prkdcscid/Jnarl	NLAC

[a]AB, *Abcam (Milton, Cambridge, United Kingdom)*; AS, *AnaSpec, Inc. (Fremont, CA, United States)*; BD, *BD Biosciences (CA, United States)*; BI, *Biological Industries (Beit-Haemek, Israel)*; BL, *BioLamina (Sundbyberg, Sweden)*; CO, *Corning (Corning, NY, United States)*; LO, *Lonza (Basel, Switzerland)*; LT, *Life Technology (CA, United States)*; MK, *Merck KGaA (Darmstadt, Germany)*; NI, *Nippi (Tokyo, Japan)*; NLAC, *National Laboratory Animal Center (Taipei, Taiwan)*; PST, *ProSpec-Tany TechnpGene Ltd. (Rehovot, Israel)*; SA, *Sigma-Aldrich (MO, United States)*; SCB, *Santa Cruz Biotechnology (TX, United States)*; SCT, *STEMCELL Technologies Inc. (BC, Canada)*; TFS, *Thermo Fisher Scientific Inc. (MA, United States)*.

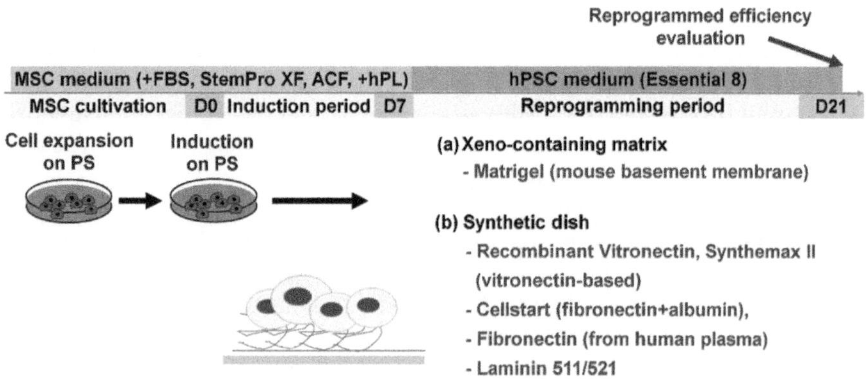

FIG. 7

Timeline of the reprogramming process of the tissue cells into hiPSCs. The tissue cells were reprogrammed on several matrices utilizing a Sendai virus vector for pluripotent gene transfection (Sung et al., 2019a).

5. Procedure

5.1 Generation of hiPSCs using Sendai virus transfection kit

Reprogramming of human amniotic fluid stem (hAFS) cells into hiPSCs was performed using CytoTune^{R}-iPS 2.0 Sendai Reprogramming Kit where this kit contains three Sendai virus-based reprogramming vectors. Fig. 7 shows schematic process of the reprogramming. The reprogramming process under xeno-free conditions was done as the following process according to the manufacturer's guidance (Higuchi et al., 2014a; Sung et al., 2019a).

(1) Day 0: Preparation of hAFS cells for gene transfection

Human AFS cells were utilized for the reprogramming process of the cells into hiPSCs.

One set of CytoTune^{R} 2.0 Sendai virus tubes were taken from refrigerator at −80.0°C. Each tube was thawed by inserting the bottom of the tube in a warm water bath at 37°C for 10 s, and then the tube was removed from the water bath and left at the place at 20–28°C. Then, the tube was lightly centrifuged and located on ice immediately. The volume of virus utilized for the transfection was estimated from the following calculation:

$$\text{Volume of virus } (\mu\text{L}) = \frac{\text{MOI}\left(\frac{\text{CIU}}{\text{cell}}\right) \times \text{No.of cells}}{\text{Titer of virus}\left(\frac{\text{CIU}}{\text{mL}}\right)} \quad (1)$$

The multiplicity of infection (MOI) was 3, 5, and 5 for hKlf4 MOI, hc-Myc MOI, and KOS MOI, respectively.

The estimated amount of each of the three CytoTune^{R} 2.0 Sendai viruses was inserted into 1.0 mL of media (ACF media, StemPro XF media, or 20%-FBS media) that was heated to 37°C. The hAFS cell media inserting the cells in the 12-well polystyrene (PS) plates were removed, and the reprogramming virus mixture was inserted into each well of PS plates including the cells. The cells were cultivated overnight in a CO_2 incubator at 37°C.

(2) Day 1–6: Cell cultivation

After 1 day of transfection, the media were changed with fresh media (ACF media, StemPro XF media, or 20%-FBS media). Then, the cells were cultivated for 6 days, and the media was changed with fresh media once every 2 days.

(3) Day 7: Shifting of the transfected cells into biomaterials

The cells were taken out from the plates utilizing 0.05% EDTA solution (0.5 mL) and left at room temperature. Once the cells were rounded up (1–3 min later), 2.0 mL of hAFS cell media was inserted into each well, and the cells were gathered in conical centrifuge tubes. The cells were centrifuged at 200 × g for 4 min, and the cells were suspended in an appropriate amount of hAFS cell media. The cells were inoculated on several different cell cultivation biomaterials and cultivated in a CO_2 incubator at 37°C. The cell cultivation biomaterials utilized in this study were ECM-immobilized 12-well PS dishes (Recombinant VN-coated dishes, Synthemax II-coated dishes, LN-521-coated dishes, and Matrigel-coated dishes) (Higuchi et al., 2014a; Sung et al., 2019a).

(4) Day 8 to 80: hiPSC generation and cultivation

On day 8, the media was changed to hPSC media (Essential 8), and the media was changed with fresh media once every day. From day 8, the cell morphologies were observed every other day under a phase inverted microscope to evaluate for the appearance of cell clumps, which indicate reprogrammed hiPSCs. After 3 weeks of transfection, the colonies were shifted to new plates that were grown to an appropriate size. When the colonies were ready to move, the live staining of cells using alkaline phosphatase or Tra-1-60 was processed for the selection of reprogrammed colonies.

We manually picked up the undifferentiated hiPSC colonies and transferred hiPSC colonies into new dishes for further usage. The effect of the cell cultivation biomaterials on the success rate of reprogramming was studied on day 28 (Higuchi et al., 2014a; Sung et al., 2019a).

5.2 Induction of hiPSC differentiation into cardiomyocytes

We followed the protocol developed by Bhattacharya and his colleagues (Bhattacharya et al., 2014), because we can differentiate hPSCs into cardiomyocytes with high reproducibility (typically 100%). On day −4, hPSCs were inoculated into ECM-coated plates, and the cells were then cultivated in E8 media until day 0, replacing the E8 cultivation media once every day. On day 0, the hPSCs became to be approximately 80%–90% confluence. The E8 cultivation media was changed with

RPMI 1640 media containing GSK-3β inhibitor (6 μM CHIR99021) and 2 wt% B27 minus insulin. The cells were cultured for 2 more days. On day 2, the media were changed with RPMI 1640 media containing 2 wt% B27 minus insulin, and the cells were cultivated for 2 more days. On day 4, the media were replaced with RPMI 1640 media containing Wnt inhibitor (5 μM IWR-1) and 2 wt% B27 minus insulin, and the cells were cultivated for 2 more days. On day 6, the media were replaced with RPMI 1640 media containing 2 wt% B27 minus insulin, and the cells were cultivated for 1 more day. On day 7, the media were replaced with RPMI 1640 media containing 2 wt% B27. The cells were cultured until day 14, during which the media were replaced with fresh media once every other day (Sung et al., 2019b).

5.3 Induction of hiPSC differentiation into MSCs

We followed the protocol developed by Li and his colleagues (Li et al., 2018).

1. Day −1: Preparation of the cells for differentiation into MSCs
   Human iPSC were inoculated on six well plates immobilized with several ECMs at about 5×10^4 cells/well in E8 media at 5% CO_2 and 37°C, and cultivated overnight.
2. Day 0: Differentiation of hiPSCs into trophoblast
   The cell culture medium in the plates was removed and the plates were washed with PBS twice. Subsequently, E6 medium (E8 medium without TGF-β1 and FGF-2) containing 1 μM A83-01 and 10 ng/mL BMP4 was added into the plate (2 mL/well). The cells were cultured at 37°C and 5% CO_2.
3. Day 3–70: Differentiation into MSCs
   The cells were cleaned with PBS once. Subsequently, the cells were cultivated in MSC medium (αMEM media containing 1 × nonessential amino acids, L-glutamine and 20% FBS) at 2 mL/well. The media were replaced with fresh media once every 2 days. The cells were passaged when the cells reached 78%–88% confluence, which took approximately 5–7 days in general. Typically, hiPSC-derived MSCs are possible to repeat passages until 10 passages.

Expression of MSC markers, CD105, CD 90, CD73, CD44, and CD34 will be analyzed using flow cytometry.

6. Conclusion

Introduction of small and large molecules can target signaling transduction pathways, which affects the reprogramming, epigenic changes, apoptosis, cell differentiation, and DNA replication of the cells into iPSCs. Some specific small bioactive molecules can successfully increase the reprogramming efficiencies of human tissue cells into iPSCs. Several feeder-free and xeno-free culture systems are also commercially available. For example, hPSCs can be cultured on (1) VN-coated plates in Essential 8 (E8) or TeSR2 media; (2) Synthemax (an oligopeptide derived from

VN)-coated plates in E8, StemPro, or TeSR2 media; (3) StemAdhere-coated plates in E8 or TeSR2 media; or (4) Poly-D-lysine-coated plates in E8 or TeSR2 containing Y27632. However, not all hiPSCs and hESCs are able to proliferate with their pluripotency in each of the above serum-free cultivation conditions. Further development of coating biomaterials and cultivation matrices, including cell culture bags, nanofibers, microcarriers, and hydrogels is necessary to be developed for the long-term cultivation of hiPSCs and hESCs as well as for the development of inexpensive cultivation media.

References

Bhattacharya, S., Burridge, P., Kropp, E., Chuppa, S., Kwok, W.-M., Wu, J., Boheler, K., Gundry, R., 2014. High efficiency differentiation of human pluripotent stem cells to cardiomyocytes and characterization by flow cytometry. J. Vis. Exp. 91. https://doi.org/10.3791/52010.

Braam, S., Zeinstra, L., Litjens, S., Oostwaard, D., Brink, S., Van Laake, L., Lebrin, F., Kats, P., Hochstenbach, R., Passier, R., Sonnenberg, A., Mummery, C., 2008. Recombinant vitronectin is a functionally defined substrate that supports human embryonic stem cell self-renewal via αVβ5 integrin. Stem Cells 26, 2257–2265. https://doi.org/10.1634/stemcells.2008-0291.

Fu, X., Toh, W.S., Liu, H., Lu, K., Li, M., Cao, T., 2011. Establishment of clinically compliant human embryonic stem cells in an autologous feeder-free system. Tissue Eng. Part C Methods 17, 927–937. https://doi.org/10.1089/ten.TEC.2010.0735.

Giorgetti, A., Montserrat, N., Aasen, T., Gonzalez Grassi, F., Rodríguez-Pizà, I., Vassena, R., Raya, A., Boue, S., Barrero, M., Aran, B., Torrabadella, M., Veiga, A., Izpisua Belmonte, J.C., 2009. Generation of induced pluripotent stem cells from human cord blood using OCT4 and SOX2. Cell Stem Cell 5, 353–357. https://doi.org/10.1016/j.stem.2009.09.008.

Higuchi, A., Ling, Q.-D., Ko, Y.-A., Chang, Y., Umezawa, A., 2011. Biomaterials for the feeder-free culture of human embryonic stem cells and induced pluripotent stem cells. Chem. Rev. 111, 3021–3035. https://doi.org/10.1021/cr1003612.

Higuchi, A., Lin, F., Cheng, Y.-K., Kao, T.-C., Kumar, S., Ling, Q.-D., Hou, C.-H., Chen, D.-C., Hsu, S.-T., Wu, G.-J., 2014a. Preparation of induced pluripotent stem cells on dishes grafted on oligopeptide under feeder-free conditions. J. Taiwan Inst. Chem. Eng. 45, 295–301. https://doi.org/10.1016/j.jtice.2013.06.022.

Higuchi, A., Ling, Q.-D., Kumar, S., Munusamy, M., Alarfajj, A., Umezawa, A., Wu, G.-J., 2014b. Design of polymeric materials for culturing human pluripotent stem cells: progress toward feeder-free and xeno-free culturing. Prog. Polym. Sci. 39. https://doi.org/10.1016/j.progpolymsci.2014.01.002.

Higuchi, A., Ling, Q.-D., Kumar, S., Murugan, A., Alarfaj, A., Chang, Y., Kao, S.-H., Lin, K.-C., Wang, H.-C., Umezawa, A., 2014c. Generation of pluripotent stem cells without the use of genetic material. Lab. Investig. 95. https://doi.org/10.1038/labinvest.2014.132.

Higuchi, A., Kumar, S., Benelli, G., Alarfaj, A., Murugan, A., Umezawa, A., Murugan, K., 2017. Stem cell therapies for reversing vision loss. Trends Biotechnol. 35. https://doi.org/10.1016/j.tibtech.2017.06.016.

Higuchi, A., Suresh Kumar, S., Benelli, G., Ling, Q.-D., Li, H.-F., Alarfaj, A.A., Munusamy, M.A., Sung, T.-C., Chang, Y., Murugan, K., 2019. Biomaterials used in stem cell therapy for spinal cord injury. Prog. Mater. Sci. 103, 374–424. https://doi.org/10.1016/j.pmatsci.2019.02.002.

Ilic, D., Stephenson, E., Wood, V., Jacquet, L., Stevenson, D., Petrova, A., Kadeva, N., Codognotto, S., Patel, H., Semple, M., Cornwell, G., Ogilvie, C., Braude, P., 2012. Derivation and feeder-free propagation of human embryonic stem cells under xeno-free conditions. Cytotherapy 14, 122–128. https://doi.org/10.3109/14653249.2011.623692.

Kaji, K., Norrby, K., Paca, A., Mileikovsky, M., Mohseni, P., Woltjen, K., 2009. Virus-free induction of pluripotency and subsequent excision of reprogramming factors. Nature 458, 771–775. https://doi.org/10.1038/nature07864.

Kim, J.B., Greber, B., Araúzo-Bravo, M., Meyer, J., Park, K.I., Zaehres, H., Schöler, H., 2009. Direct reprogramming of human neural stem cells by OCT4. Nature 461, 643–649. https://doi.org/10.1038/nature08436.

Lee, J., Xia, Y., Son, M.-Y., Jin, G., Seol, B., Kim, M.-J., Son, M., Do, M., Lee, M., Kim, D., Lee, K., Cho, Y., 2012. A novel small molecule facilitates the reprogramming of human somatic cells into a pluripotent state and supports the maintenance of an undifferentiated state of human pluripotent stem cells. Angew. Chem. Int. Ed. Eng. 51. https://doi.org/10.1002/anie.201206691.

Li, W., Zhou, H., Abujarour, R., Zhu, S., Joo, J., Lin, T., Hao, E., Schöler, H., Hayek, A., Ding, S., 2009. Generation of human induced pluripotent stem cells in the absence of exogenous Sox2. Stem Cells 27, 2992–3000. https://doi.org/10.1002/stem.240.

Li, E., Zhang, Z., Jiang, B., Yan, L., Park, J., 2018. Generation of mesenchymal stem cells from human embryonic stem cells in a complete serum-free condition. Int. J. Biol. Sci. 14, 1901–1909. https://doi.org/10.7150/ijbs.25306.

Meng, G., Liu, S., Rancourt, D., 2011. Synergistic effect of medium, matrix, and exogenous factors on the adhesion and growth of human pluripotent stem cells under defined, Xeno-free conditions. Stem Cells Dev. 21, 2036–2048. https://doi.org/10.1089/scd.2011.0489.

Miyazaki, T., Futaki, S., Hasegawa, K., Kawasaki, M., Sanzen, N., Hayashi, M., Kawase, E., Sekiguchi, K., Nakatsuji, N., Suemori, H., 2008. Recombinant human laminin isoforms can support the undifferentiated growth of human embryonic stem cells. Biochem. Biophys. Res. Commun. 375, 27–32. https://doi.org/10.1016/j.bbrc.2008.07.111.

Rodin, S., Domogatskaya, A., Ström, S., Hansson, E., Chien, K., Inzunza, J., Hovatta, O., Tryggvason, K., 2010. Long-term self-renewal of human pluripotent stem cells on human recombinant laminin-511. Nat. Biotechnol. 28, 611–615. https://doi.org/10.1038/nbt.1620.

Soldner, F., Hockemeyer, D., Beard, C., Gao, Q., Bell, G., Cook, E., Hargus, G., Blak, A., Cooper, O., Mitalipova, M., Isacson, O., Jaenisch, R., 2009. Parkinson's disease patient-derived induced pluripotent stem cells free of viral reprogramming factors. Cell 136, 964–977. https://doi.org/10.1016/j.cell.2009.02.013.

Sung, T.-C., Li, H.-F., Higuchi, A., Kumar, S., Ling, Q.-D., Wu, Y.-W., Burnouf, T., Nasu, M., Umezawa, A., Lee, K.-F., Wang, H.-C., Chang, Y., Hsu, S.-T., 2019a. Effect of cell culture biomaterials for completely xeno-free generation of human induced pluripotent stem cells. Biomaterials 119638. https://doi.org/10.1016/j.biomaterials.2019.119638.

Sung, T.-C., Liu, C.-H., Huang, W.-L., Lee, Y.-C., Kumar, S., Chang, Y., Ling, Q.-D., Hsu, S.-T., Higuchi, A., 2019b. Efficient differentiation of human ES and iPS cells into cardiomyocytes on biomaterials under xeno-free conditions. Biomater. Sci. 7. https://doi.org/10.1039/C9BM00817A.

Takahashi, K., Tanabe, K., Ohnuki, M., Narita, M., Ichisaka, T., Tomoda, K., Yamanaka, S., 2007. Induction of pluripotent stem cells from adult human fibroblasts by defined factors. Cell 131, 861–872. https://doi.org/10.1016/j.cell.2007.11.019.

Vodyanik, M., Smuga-Otto, K., Antosiewicz-Bourget, J., Frane, J., Tian, S., Nie, J., Jonsdottir, G., Ruotti, V., Stewart, R., Slukvin, I., Thomson, J., 2008. Induced pluripotent stem cell lines derived from human somatic cells. Science 318, 1917–1920. https://doi.org/10.1126/science.1151526.

Wang, Q., Xinxiu, X., Li, J., Liu, J., Gu, H., Zhang, R., Chen, J., Kuang, Y., Fei, J., Jiang, C., Wang, P., Pei, D., Ding, S., Xie, X., 2011. Lithium, an anti-psychotic drug, greatly enhances the generation of induced pluripotent stem cells. Cell Res. 21, 1424–1435. https://doi.org/10.1038/cr.2011.108.

Woltjen, K., Michael, I., Mohseni, P., Desai, R., Mileikovsky, M., Hämäläinen, R., Cowling, R., Wang, W., Liu, P., Gertsenstein, M., Kaji, K., Sung, H.-K., Nagy, A., 2009. PiggyBac transposition reprograms fibroblasts to induced pluripotent stem cells. Nature 458, 766–770. https://doi.org/10.1038/nature07863.

Yap, L., Li, J., Phang, I., Ong, L., Ow, J., Goh, J., Nurcombe, V., Hobley, J., Choo, A., Oh, S., Cool, S., Birch, W., 2011. Defining a threshold surface density of Vitronectin for the stable expansion of human embryonic stem cells. Tissue Eng. Part C Methods 17, 193–207. https://doi.org/10.1089/ten.TEC.2010.0328.

Zhu, S., Li, W., Zhou, H., Wei, W., Ambasudhan, R., Lin, T., Kim, J., Zhang, K., Ding, S., 2010. Reprogramming of human primary somatic cells by OCT4 and chemical compounds. Cell Stem Cell 7, 651–655. https://doi.org/10.1016/j.stem.2010.11.015.

CHAPTER

Confirmation of stem cells pluripotency via teratoma formation in an animal model☆

7

Seoh Wei Teh[a], Pooi Ling Mok[b], and Suresh Kumar Subbiah[c]

[a]*Department of Orthopaedics, Faculty of Medicine and Health Sciences, Universiti Putra Malaysia, Serdang, Selangor, Malaysia,* [b]*Department of Biomedical Science, Faculty of Medicine and Health Sciences, Universiti Putra Malaysia, Serdang, Selangor, Malaysia,* [c]*Centre for Materials Engineering and Regenerative Medicine, Bharath Institute of Higher Education and Research, Chennai, Tamil Nadu, India*

1. Establishment of teratoma in small animal model

1.1 Introduction

Stem cells are pluripotent, capable of giving rise to cells from three germ layers, namely endoderm, mesoderm, and ectoderm (Chen et al., 2015). Stem cells can be xeno-transplanted into severe combined immunodeficiency (SCID) mice via subcutaneous injection to generate teratomas (Higuchi et al., 2011). About 2 months later, the resulting teratomas can be harvested and dissected to analyze the manifestation of cells that develop from the three germ layers. For instance, enteron is derived from the endoderm; osteoblasts and chondrocytes are derived from mesoderm; while neurons are derived from the ectoderm (Liu et al., 2014). The presence of these cells can be determined by simple staining with Hematoxylin and Eosin (H&E) (Zhou et al., 2009). Formation of teratomas in small animal model showing cells from multigerm lineages of endodermal, mesodermal, and ectodermal will support the notion that the stem cells are capable to derive into cells of all the three lineages in vivo and, therefore, could potentially form almost all types of cells in the body.

☆Important note: Animal research ethical approval should be obtained from the Institutional Review Board.

Stem Cell Laboratory Techniques. https://doi.org/10.1016/B978-0-12-823729-8.00010-5

1.2 Materials

Stem cells in single cell suspension	5×10^6 cells
Complete growth media: Dulbecco's Modified Eagle Medium/Ham's F-12 (DMEM/F12) with 10% fetal bovine serum (FBS)	10 mL
0.25% Trypsin-ethylenediaminetetraacetic acid (EDTA)	3.0 mL
80 μg/mL Matrigel-DMEM/F12 (1.0 mg Matrigel in 12 mL of ice-cold DMEM-F12)	10 mL
Matrigel matrix high concentration (Cat No: CLS354262; Corning; New York, United States)	
Severe combined immunodeficiency (SCID) mice, 6–8 weeks old	
Ketamine (Cat No: K2753; Sigma Aldrich; Steinheim, Germany)	
Xylazine (Cat No: X1126; Sigma Aldrich; Steinheim, Germany)	

1.3 Protocols

1. Harvest cells from culture surface using 0.25% trypsin-EDTA (1 mL for a 25 cm^2 flask) or cell scraper (Cat No: CLS3010; Corning; New York, United States).
2. If trypsin-EDTA is used, add 2–3 times the volume of complete growth media containing FBS to the cells to stop the trypsinization.
3. Transfer the cells into a 15 mL centrifuge tube and centrifuge the cells at $200 \times g$ for 5 min.
4. Discard the supernatant.
5. Resuspend the pellet in 80 μg/mL Matrigel-DMEM/F12.

   Note: All media and cultureware in contact with Matrigel must be ice-cold to avoid solidification of Matrigel.
6. Anesthetize the mouse with ketamine (100 mg/kg) and xylazine (10 mg/kg) through intraperitoneal injection.
7. Inject 5×10^6 cells into the SCID mouse in the subcutaneous layer of the left flank (Fig. 1).
8. The animal should be monitored and provided with distilled water and rodent pellets (Zeigler; Pennsylvania, United States) ad libitum.

2. Histological examination on formed teratoma

2.1 Introduction

Teratomas generated from stem cells injected into animal model contain cells derived from the three germ layers, namely endoderm, mesoderm, and ectoderm (Miyawaki et al., 2017). Endoderm is the innermost of the three primary germ layers of an embryo. It develops into the respiratory tract and gastrointestinal tract. Mesoderm is the middle germ layer, which develops into muscles, bones, cartilage, connective tissue, urogenital system, and circulatory system. Ectoderm is the outermost of the three primary germ layers, developing the sense organs and nervous system.

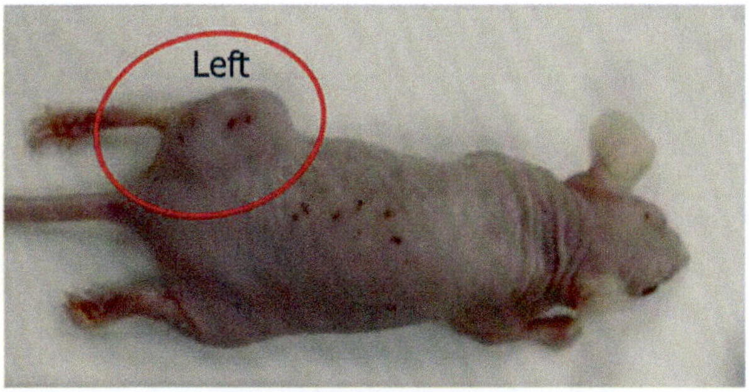

FIG. 1

Teratomas *(red circle)* formed in SCID mouse after 6–8weeks.

Adapted from Higuchi, A., Kao, S.H., Ling, Q.D., Chen, Y.M., Li, H.F., Alarfaj, A.A., Hsu, S.T., 2015. Long-term xeno-free culture of human pluripotent stem cells on hydrogels with optimal elasticity. Sci. Rep. 5, 18136.

Epidermises including the skin, hair and nails, linings of the mouth, anus, nostrils, sweat glands, and tooth enamel are also derived from ectoderm (Lee et al., 2014).

The histological structure of cells derived from the three germ layers can be determined by H&E staining of tissue sections of the teratoma formed in animal model (Bulic-Jakus et al., 2016). Following dissection of teratoma, the standard steps for teratoma processing are fixation in paraformaldehyde, embedding in paraffin wax, and sectioning using a microtome to produce microscopic slides that can be stained and viewed under a microscope. H&E staining confers contrast of color to the cellular structures on the teratoma sections. Hematoxylin stains the cell nuclei blue or dark-purple, whereas Eosin stains the cytoplasm and extracellular matrix pink. Other basophilic structures, such as calcified material and keratohyalin granules, are stained blue. Intracellular or extracellular proteins are eosinophilic structures and, therefore, are stained pink.

The digestive tract or enteron including the esophagus, stomach, small intestine, and large intestine is developed in the endoderm. The H&E staining of enteron is characterized by long and narrow mucosal folds with short and branched villi. The cell nuclei are stained blue while cytoplasm is stained pink (Fig. 2).

Bones and cartilages are developed in the mesoderm. The cell nuclei of osteoblast cells are spherical and large (Leow et al., 2015). Following H&E staining, the cell nuclei appear blue-purple. The active osteoblast cells secrete collagen type I to the bone matrix. Due to the presence of collagen, the bone matrix is stained pink by Eosin. The area next to the nucleus of osteoblast cells is the Golgi apparatus, which is not stained. On the other hand, the round-shaped chondrocytes are present in clusters called "cell nests." The matrix immediately adjacent to chondrocytes, namely territorial matrix contains large amounts of glycosaminoglycans which are basophilic (Zhou et al., 2019). Therefore, the territorial matrix is stained dark pink by Eosin. The interterritorial matrix located further away from chondrocytes, composing of fewer glycosaminoglycans, proteoglycans, type II collagen, and chondroitin sulfate is slightly stained pink by Eosin.

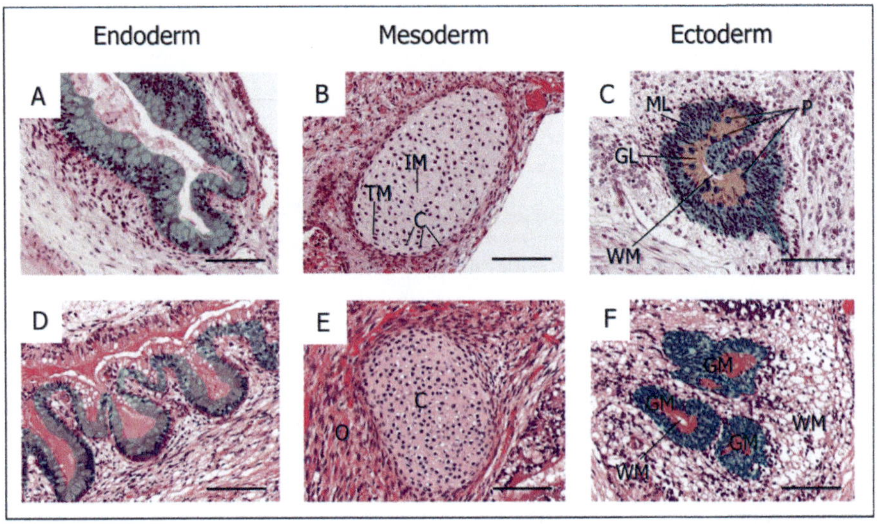

FIG. 2

H&E staining of teratoma sections showed cells originated from endoderm, mesoderm, and ectoderm. The enteron developed in the endoderm is characterized by mucosal folds with branched villi (*green color regions*) (A and D). Bones and cartilages developed in the mesoderm show *pink* matrix of chondrocytes and purple cell nuclei (B and E). The brain tissues developed in the ectoderm form some foldings of *gray matter* in the middle of *white matter*, where the three layers of *gray matter* are stained distinctly by H&E (*green color regions*, molecular layer; *yellow color regions*, granular layer; *blue color dots*, Purkinje cells), while the *white matter* is characterized by myelinated axons with *purple* nuclei and large nonstained cytoplasm (C and F). *C*, chondrocytes; *TM*, territorial matrix; *IM*, interterritorial matrix; *O*, osteoblast; *ML*, molecular layer; *GL*, granular layer; *P*, Purkinje cells; *WM*, white matter; *GM*, gray matter.

Adapted and modified from Fukusumi, H., Shofuda, T., Kanematsu, D., Yamamoto, A., Suemizu, H., Nakamura, M., 2013. Feeder-free generation and long-term culture of human induced pluripotent stem cells using pericellular matrix of decidua derived mesenchymal cells. PLoS One 8(1), e55226.

Brain tissues are developed in the ectoderm. The gray matter of the brain contains neurons and glial cells, whereas the white matter contains myelinated axons and oligodendrocytes (Adams et al., 2015). The outermost layer of the gray matter, namely molecular layer containing axons will stain pink, the thin middle layer, namely Purkinje cell layer containing large Purkinje neurons will stain purple, while the innermost layer, namely granular layer containing small neurons will stain dark purple (Genser-Nir et al., 2016). On the other hand, the presence of myelinated axons, represented by the dark purple stain in the nuclei and nonstained large cytoplasm, is the most prominent feature of white matter.

By observing the histological slides of teratoma generated from stem cells injected into animal model, the pluripotency of stem cells can be confirmed. This step is important to ensure the use of correct cell type and successful stem cell research prior to the experiments (Park et al., 2015).

2.2 Materials

10% Paraformaldehyde in phosphate-buffered saline (PBS) (Cat No.: 158127; Sigma-Aldrich; Steinheim, Germany)
70%, 80%, 95%, and 100% ethanol solution (Sigma-Aldrich; Steinheim, Germany)
Xylene (Cat No: 534056; Sigma-Aldrich; Steinheim, Germany)
Hematoxylin (Cat No: GHS3128; Sigma-Aldrich; Steinheim, Germany)
Eosin (Cat No: 318906; Sigma-Aldrich; Steinheim, Germany)
1% Acid alcohol (Cat No: SIG/A3957; Sigma-Aldrich; Steinheim, Germany)
Mounting medium (Cat No: 06522; Sigma-Aldrich; Steinheim, Germany)
Warm water bath
Microtome
Hot plate
Chiller plate
Scalpel
Glass slides

2.3 Protocols

2.3.1 Harvest and processing of teratoma

1. After 6–8 weeks, observe for the presence of teratoma on the mouse.
2. Anesthetize the mouse with ketamine (100 mg/kg) and xylazine (10 mg/kg) through intraperitoneal injection.
3. Euthanize the mouse by cervical dislocation and dissect the teratoma using a set of dissection kit.
4. Immerse the teratoma in 10% paraformaldehyde in PBS in a 50 mL polypropylene tube for 4 h.
5. Immerse the teratoma in 70%, 80%, 95%, and 100% ethanol for 1 h each for dehydration.
6. Immerse the teratoma in xylene for 2 h.
7. Immerse the teratoma in molten paraffin wax of 50–65°C for 2 h with two changes until solidify.

   Note: Processing of teratoma, i.e., steps 4–6 can be done in an automated processor (Fig. 3B).

2.3.2 Embedding of teratoma tissues

1. Use an embedding mold of standard size to contain the teratoma tissues.
2. Heat the paraffin wax till it is liquefied and poured it into the mold.
3. Place the teratoma tissue into the mold on a hot plate and adjust its position so that the desired tissue plane of investigation can be obtained during sectioning (Fig. 3C).
4. Transfer the mold to a small cold plate to allow solidification of the wax. It is important to maintain the desired orientation of the tissue during this step.
5. Place a cassette base on top of the mold and add more molten wax to fill the cassette.
6. Transfer the mold onto a large chiller plate and left to solidify.

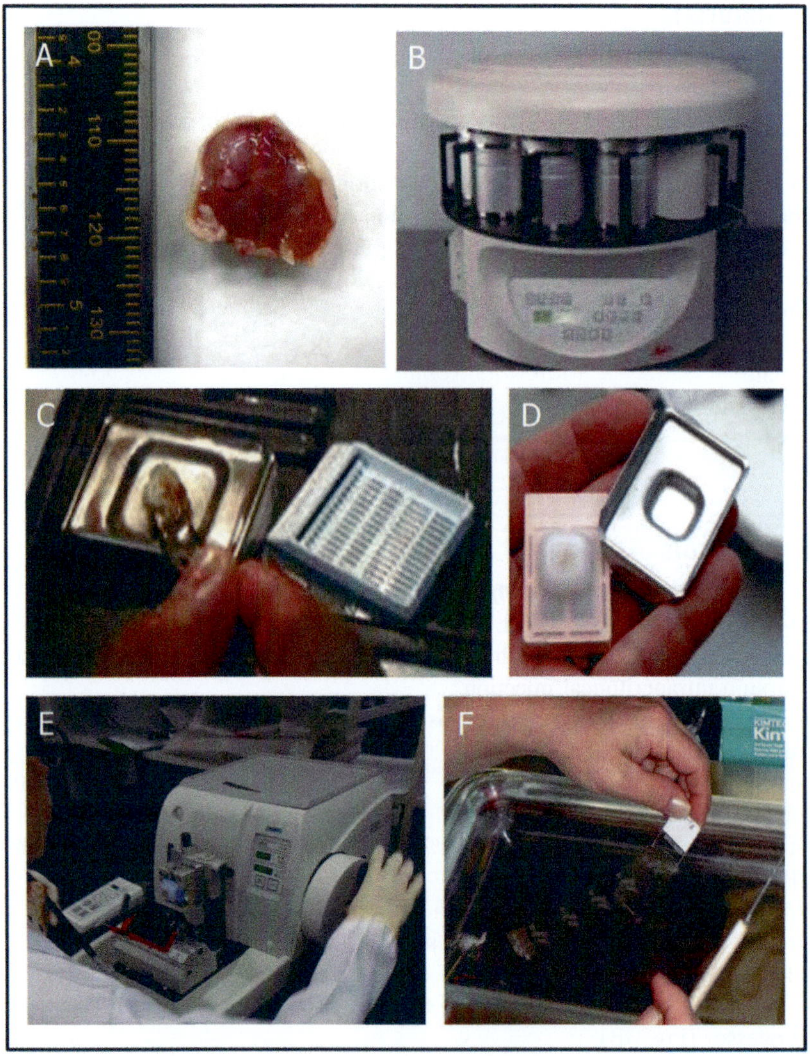

FIG. 3

Procedures of histological examination on formed teratoma. (A) Dissection of teratoma from the subcutaneous layer of SCID mouse. (B) Processing of teratoma tissue in an automated processor. (C) Placement of teratoma tissue in an embedding mold on a hot plate. (D) Paraffin-embedded teratoma tissue sample in an embedding mold. (E) Sectioning of paraffin-embedded teratoma using a microtome. (F) Transfer of teratoma tissue section onto a glass slide by placing a glass slide under the water to lift the floating tissue section up.

(A) Adapted from Higuchi, A., Kao, S.H., Ling, Q.D., Chen, Y.M., Li, H.F., Alarfaj, A.A., Hsu, S.T., 2015. Long-term xeno-free culture of human pluripotent stem cells on hydrogels with optimal elasticity. Sci. Rep. 5, 18136.

2.3.3 Sectioning of teratoma tissues

1. Take out paraffin-embedded teratoma from the mold (Fig. 3D).
2. Remove excess wax from the wax block of teratoma using a scalpel, while maintaining a margin of at least 2.0mm around the sample.
3. Set the microtome to section the wax block at 12 μm to remove all the excess wax that cover on top of the specimen.
4. Once the desired tissue plane of investigation is exposed, section the paraffin-embedded teratomas at 4.0 μm thickness using a microtome (Fig. 3E).

   Note: Refrigerate the wax block of teratoma at 4°C until the block is being sectioned. Heat that is produced from friction after sectioning for a long time will affect the quality of the ribbon produced.
5. Carefully transfer the section ribbon to allow floating on warm water. This is important to avoid the folding of the thin tissue sections.
6. Transfer the sections onto a glass slide by placing a glass slide under the water to lift the tissue section up (Fig. 3F).
7. Place the slides on a hot plate at 30–35°C for 5 min to allow better adherence before staining.

2.3.4 H&E staining of teratoma slides

1. Immerse the slides in in xylene for 3 min.
2. Immerse the slides in 100%, 95%, and 70% ethanol solution for 3 min each for rehydration.
3. Rinse the slides under running tap water for 3 min.
4. Immerse the slides in Hematoxylin for 5 min and rinse under running tap water for 3 min.
5. Immerse the slides in 1% acid alcohol for 3 s and rinse under running tap water for 1 min.
6. Immerse the slides thrice in Eosin for 1 min and rinse under running tap water for 10 s.
7. Immerse the slides in 70%, 95%, and 100% ethanol solution for 1 min each for dehydration.
8. Immerse the slides in in xylene for 3 min for clearing.
9. Mount a cover slip on each slide with mounting medium.
10. View the slides under a light microscope.

3. Conclusion

Teratoma formation is important in indicating the main hallmark of stem cells, its multilineage differentiation potential. When being injected into animal models of immunodeficiency, stem cells will form teratomas, which consist of cells derived from the three germ layers. This supports the hallmark of stem cells that they could differentiate into cells of endoderm, mesoderm, and ectoderm in vivo. The test cells can be confirmed as pluripotent stem cells if only the H&E stained section of the teratomas showed the presence of all these three types of cells. This chapter has delineated the procedures of teratoma establishment in a small animal model, teratoma harvesting, and H&E staining of teratoma sections.

Abbreviations

DMEM/F-12	Dulbecco's Modified Eagle Medium/Ham's F-12
EDTA	ethylenediaminetetraacetic acid
FBS	fetal bovine serum
H&E	hematoxylin and eosin
PBS	phosphate-buffered saline
SCID	severe combined immunodeficiency

References

Adams, K.L., Rousso, D.L., Umbach, J.A., Novitch, B.G., 2015. Foxp1-mediated programming of limb-innervating motor neurons from mouse and human embryonic stem cells. Nat. Commun. 6 (1), 1–16.

Bulic-Jakus, F., Katusic Bojanac, A., Juric-Lekic, G., Vlahovic, M., Sincic, N., 2016. Teratoma: from spontaneous tumors to the pluripotency/malignancy assay. Wiley Interdiscip. Rev. Dev. Biol. 5 (2), 186–209.

Chen, H., Aksoy, I., Gonnot, F., Osteil, P., Aubry, M., Hamela, C., Mure, M., 2015. Reinforcement of STAT3 activity reprogrammes human embryonic stem cells to naive-like pluripotency. Nat. Commun. 6, 7095.

Genser-Nir, M., Daniliuc, S., Tevrovsky, M., Brill, R.H., Miropolski, Y.Y., Fiorentini, D., 2016. A novel set of serum-free, xeno-free differentiation media for adipogenesis, osteogenesis and chondrogenesis of human mesenchymal stem cells from various tissue sources. Cytotherapy 18 (6), S115.

Higuchi, A., Ling, Q.D., Ko, Y.A., Chang, Y., Umezawa, A., 2011. Biomaterials for the feeder-free culture of human embryonic stem cells and induced pluripotent stem cells. Chem. Rev. 111 (5), 3021–3035.

Lee, J.H., Lee, J.B., Shapovalova, Z., Fiebig-Comyn, A., Mitchell, R.R., Laronde, S., Bhatia, M., 2014. Somatic transcriptome priming gates lineage-specific differentiation potential of human-induced pluripotent stem cell states. Nat. Commun. 5 (1), 1–13.

Leow, S.N., Luu, C.D., Nizam, M.H., Mok, P.L., Ruhaslizan, R., Wong, H.S., Bastion, M.L.C., 2015. Safety and efficacy of human Wharton's jelly-derived mesenchymal stem cells therapy for retinal degeneration. PLoS One 10 (6).

Liu, L., Yoshioka, M., Nakajima, M., Ogasawara, A., Liu, J., Hasegawa, K., Chen, Y., 2014. Nanofibrous gelatin substrates for long-term expansion of human pluripotent stem cells. Biomaterials 35 (24), 6259–6267.

Miyawaki, S., Okada, Y., Okano, H., Miura, K., 2017. Teratoma formation assay for assessing pluripotency and tumorigenicity of pluripotent stem cells. Bio Protoc. 7 (16).

Park, H.J., Yang, K., Kim, M.J., Jang, J., Lee, M., Kim, D.W., Cho, S.W., 2015. Bio-inspired oligovitronectin-grafted surface for enhanced self-renewal and long-term maintenance of human pluripotent stem cells under feeder-free conditions. Biomaterials 50, 127–139.

Zhou, H., Wu, S., Joo, J.Y., Zhu, S., Han, D.W., Lin, T., Siuzdak, G., 2009. Generation of induced pluripotent stem cells using recombinant proteins. Cell Stem Cell 4 (5), 381–384.

Zhou, S., Chen, S., Jiang, Q., Pei, M., 2019. Determinants of stem cell lineage differentiation toward chondrogenesis versus adipogenesis. Cell. Mol. Life Sci. 76 (9), 1653–1680.

CHAPTER

Gene editing for stem cells by CRISPR-Cas9

8

Seoh Wei Teh[a], Hernandez Maradiaga Oscar David[b], Avin Ee-Hwan Koh[c], Suresh Kumar Subbiah[d], and Pooi Ling Mok[c]

[a]*Department of Orthopaedics, Faculty of Medicine and Health Sciences, Universiti Putra Malaysia, Serdang, Selangor, Malaysia,* [b]*Department of Medical Microbiology, Faculty of Medicine and Health Sciences, Universiti Putra Malaysia, Serdang, Selangor, Malaysia,* [c]*Department of Biomedical Science, Faculty of Medicine and Health Sciences, Universiti Putra Malaysia, Serdang, Selangor, Malaysia,* [d]*Centre for Materials Engineering and Regenerative Medicine, Bharath Institute of Higher Education and Research, Chennai, Tamil Nadu, India*

1. Introduction

1.1 CRISPR background

Clustered regularly interspaced short palindromic repeats (CRISPR)-associated (Cas)-related system is an immunity mechanism against viruses in bacteria and archaea (Han and She, 2017). These repeats were found accidentally in 1987 in *Escherichia coli*, specifically in a nucleotide sequence of the *iap* gene (Ishino et al., 1987).

The study by Mojica et al. on *Haloferax mediterranei* led to the identification for the first time of a long deoxyribonucleic acid (DNA) sequence that contained regularly spaced repeats (Mojica et al., 1993). Jansen et al. coined the acronym CRISPR in 2002 as a term that better reflected the characteristic structures of the repeats; in the same study, it was reported the identification of four CRISPR-associated genes, Cas1 to Cas4 (Jansen et al., 2002). Mojica et al. reported in 2005 that DNA from bacteriophages was contained in spacers (Mojica et al., 2005).

Three years later, the protospacer adjacent motifs (PAM) were identified as the preferential target sites for Cas endonucleases (Horvath et al., 2008). In the next decade, three major types (type I, II, and III) of CRISPR were delineated, divided into several subtypes and chimeric variants (Makarova et al., 2011); ultimately type II interference became the system used in mammalian cells which was adapted for genome-editing purposes (Wu et al., 2014).

Stem Cell Laboratory Techniques. https://doi.org/10.1016/B978-0-12-823729-8.00009-9

1.2 CRISPR-Cas9 natural pathway

Since its discovery in the bacterial immune system (Ishino et al., 1987), CRISPR-Cas9 has become an innovative tool in the genome editing field. The popularity of the technology is because of the advantages in terms of costs, precision, and easy performance (Lee et al., 2018).

In certain locations in the genome of some prokaryotes and archaea, one particular DNA sequence is repeated frequently, and between these repeats, spacers can be found in the form of specific and unique sequences. There are Cas proteins also known as cluster associated proteins located near the repeats, many of which have enzymatic activity. Despite there being several types of Cas variants, the most widely used is Cas9, an endonuclease that has high utility in the CRISPR-Cas genome editing (Cong et al., 2013). The spacers in CRISPR locations represent a historical record of DNA viruses that have previously infected the host, thus whenever viral DNA attempts to infect a cell, the host genome is modified by the incorporation of certain fragments called protospacers which become spacers in specific CRISPR sites (Rath et al., 2015).

A pre-CRISPR ribonucleic acid (RNA) is formed after the transcription of a CRISPR gene into RNA; the pre-CRISPR RNA contains both the spacers and repeats. Another RNA, the transactivating RNA (tracrRNA) then works along with the pre-CRISPR RNA to recruit enzymes to cut the pre-CRISPR RNA into a smaller sequence that results in a mature CRISPR RNA (crRNA). When the same virus infects the cell, Cas9 protein can recognize a specific PAM site, located near the crRNA sequence. If the crRNA/tracrRNA complex known as single guide RNA (sgRNA) binds to its complementary sequence near a PAM, it helps guide Cas9 to cut the double-stranded DNA of the virus at three base pairs upstream of the PAM leading to viral destruction, representing an evolutionary adaptation against DNA viral infections (Karvelis et al., 2013; Le Rhun et al., 2019; Shmakov et al., 2015) (Fig. 1).

1.3 Fundamental process of DNA repair

Gene modification can occur by taking advantage of the cellular DNA repair mechanisms. Once targeted DNA double-strand break (DSB) occurs at specific locations, the cell machinery can repair it in one of two ways: nonhomologous end joining (NHEJ) and homology-directed repair (HDR). NHEJ is the most common repair mechanism in mammalian cells (Bennardo et al., 2008; Miyaoka et al., 2016).

Repair via NHEJ can create short insertions or deletions, i.e., indels. If an indel results in a frameshift mutation for all copies of the targeted gene, that gene is considered to be knocked out (Zuo et al., 2017). Alternatively, knock-in mutation can be created by supplying an exogenous template during repair of DSB via HDR, thus allowing precise genome editing (Hsu et al., 2014) (Fig. 2).

CRISPR-Cas gene editing techniques have been explored in the study of diseases such as pulmonary diseases (Schwank et al., 2013), gastrointestinal disorders (Pankowicz et al., 2016; Yin et al., 2014), hematological diseases (Dever et al., 2016; Ou et al., 2016; Zhang et al., 2016), viral diseases (Kennedy et al., 2014; Puschnik et al., 2017; Zhen et al., 2014), vector diseases (Kistler et al., 2015;

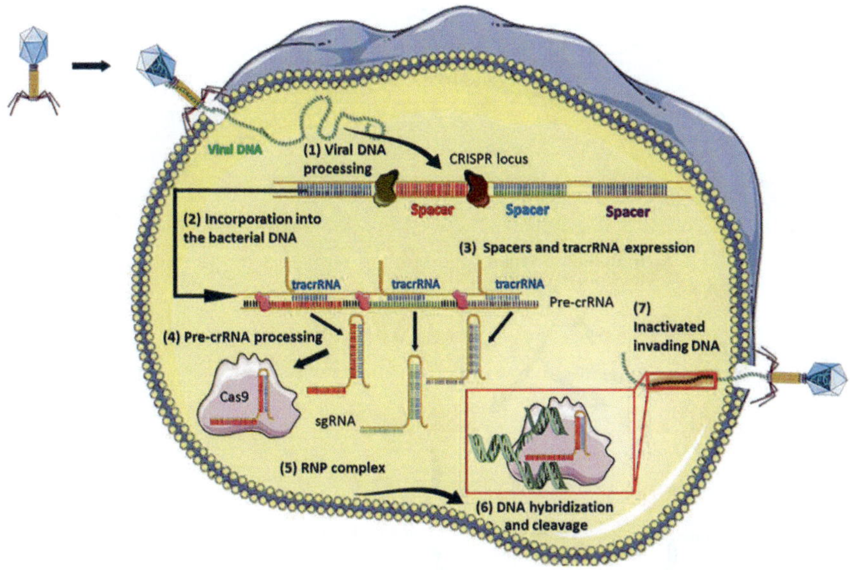

FIG. 1

A general overview of a CRISPR-Cas9 natural pathway in prokaryotes. A sgRNA guides Cas9 protein to a specific DNA sequence of the invading virus and the Cas9 protein cleave the DNA, leading to viral destruction. *RNP*, ribonucleoprotein.

Adapted and modified from Le Rhun, A., Escalera-Maurer, A., Bratovič, M., Charpentier, E., 2019. CRISPR-Cas in streptococcus pyogenes. RNA Biol. 16, 380–389.

Zhang et al., 2014), cancer (Hoellerbauer et al., 2020; Jiang et al., 2020), autoimmune and inflammatory diseases (Brunger et al., 2017; Tang et al., 2014).

The first use of CRISPR-Cas9 for gene modification in human cells, particularly the hematopoietic stem cells (HSCs), for treating transfusion-dependent thalassemia (clinicaltrials.gov; NCT03655678) or severe sickle cell disease in patients (clinicaltrials.gov; NCT03745287), has been reported in New England Journal of Medicine in 2021. The HSCs were obtained from the patients and the gRNA was designed to target on *BCL11A* erythroid-specific enhancer in the cells. The patients demonstrated high levels of allelic editing and increase of fetal hemoglobin in the blood (Frangoul et al., 2021). Fig. 3 illustrates the brief protocol to isolate blood stem cells for genetic modification from the patient.

1.4 Validation of CRISPR editing

Following CRISPR editing, the targeted genomic sequences must be analyzed to identify the presence of modifications. There are a variety of analysis tools employing different mechanisms useful for validation of CRISPR editing and determination of editing efficiency (Table 1).

Mutation detection assays such as T7 endonuclease 1 (T7E1) are simple methods used to determine on-target genome editing. Polymerase chain reaction (PCR)

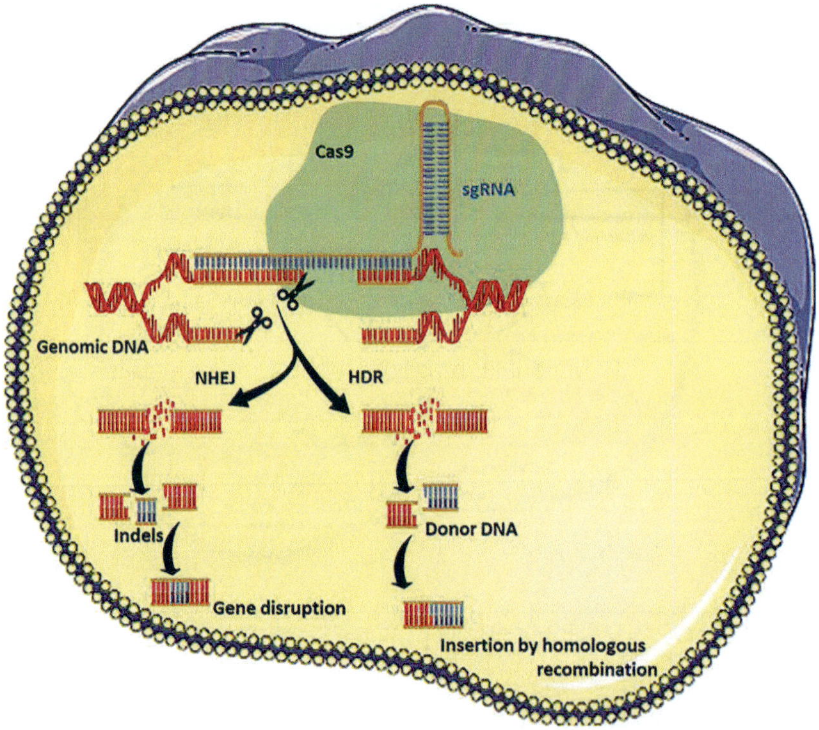

FIG. 2

Double-strand break repair mechanisms; indels are introduced during NHEJ repairs while a foreign DNA template can be inserted through HDR.

Adapted and modified from Hsu, P.D., Lander, E.S., Zhang, F., 2014. Development and applications of CRISPR-Cas9 for genome engineering. Cell 157, 1262–1278.

amplification of CRISPR-modified cells at the target genomic regions followed by denaturation and reannealing procedures results in heteroduplexed DNA formation between wild-type DNA and CRISPR-modified DNA. Mutations are then recognized by T7EI, which cleaves the mismatched DNA heteroduplexes. The generated smaller-sized DNA fragments (cleaved products) and full-length amplicons (wild-type DNA) can be visualized by agarose gel electrophoresis. T7E1 assay can be used to detect the occurrence of CRISPR-mediated genetic mutations, but it does not provide information about the generated indel sequences (Vouillot et al., 2015). Therefore, it serves as a quick assay for optimization of CRISPR editing.

CRISPR editing efficiency can be determined by reverse transcription quantitative polymerase chain reaction (RT-qPCR) through the study of relative fold change in target gene expression after CRISPR transfection. Protein assays such as western blot can also be used to evaluate the effect of gene editing on the expression of protein encoded by the gene of interest.

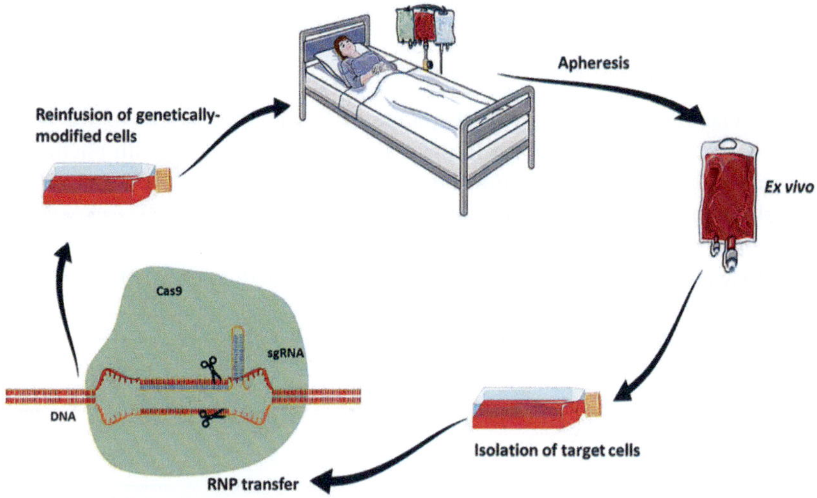

FIG. 3

Stem cell-based gene therapy. Hematopoietic stem cells isolated from patients are edited and reintroduced into the patient. Blood will be collected from patient via apheresis. Hematopoietic stem cells will then be isolated and expanded. Following CRISPR transfection on hematopoietic stem cells to modify the gene of interest, the corrected cells will then be transplanted back to the patient. *RNP*, ribonucleoprotein.

Adapted and modified from Duran, A.G., Reidell, O., Stachelscheid, H., et al., 2018. Regenerative medicine/cardiac cell therapy: pluripotent stem cells. Thorac. Cardiovasc. Surg. 7 (1), 53–62.

Table 1 Analysis methods for validation of CRISPR editing.

Method	Detect
Mutation detection assay	Indels at target sites
RT-qPCR	Change in RNA expression
Western blot	Change in protein expression
Enzyme-linked immunosorbent assay	Change in protein expression
Next generation sequencing	Overall genomic changes

The gold standard for precise analysis of CRISPR-mediated genetic modification involves deep sequencing on the region of interest by targeted next generation sequencing (NGS). Targeted NGS is an exceptionally sensitive high-throughput sequence-based method capable of detecting CRISPR edited products. Besides providing comprehensive data of the indels generated, targeted NGS allows for downstream analysis to evaluate the effect of CRISPR gene editing on the downstream signaling pathways, thereby revealing possible novel connection of a gene to a cellular function.

2. Materials

2.1 Cell culture

1. 1 cryovial containing 1×10^6 cells (human A549 lung adenocarcinoma cells)
2. 0.25% trypsin-ethylenediaminetetraacetic acid (EDTA) (1 ×) with phenol red
3. Penicillin-streptomycin mixture (PS)
4. Fetal bovine serum (FBS), heat-inactivated
5. Complete growth medium
   - Roswell Park Memorial Institute (RPMI) 1640 Medium, 5% FBS, and 1% PS, stored at 4°C
6. Phosphate-buffered saline (PBS)

2.2 Transfection

1. Synthetic gRNA*
2. Opti-Minimal Essential Medium (Opti-MEM) medium (Gibco; Massachusetts, United States)
3. TrueCut Cas9 protein v2 (Invitrogen; Massachusetts, United States)
4. Lipofectamine Cas9 Plus reagent (Invitrogen; Massachusetts, United States)
5. Lipofectamine CRISPRMAX Cas9 transfection reagent (Invitrogen; Massachusetts, United States)

   *The sequence of sgRNA targeting to different genes can be generated from the following website: https://sg.idtdna.com/site/order/designtool/index/CRISPR_CUSTOM

2.3 Validation of CRISPR editing

2.3.1 T7 endonuclease 1 assay

1. DNA extraction kit
2. Alt-R Genome Editing Detection Kit
3. Alt-R hypoxanthine phosphoribosyltransferase (HPRT) positive control
4. PCR kit
5. Nuclease-free water
6. PCR primers specific for target gene
7. 1 × PBS
8. DNA extraction kit
9. Thermal cycler
10. Agarose gel for electrophoresis

2.3.2 RT-qPCR

1. 0.25% trypsin-EDTA (1 ×) with phenol red
2. RNeasy Mini Kit (Qiagen; Maryland, United States)

3. 70% ethanol
4. SuperScript IV First-Strand Synthesis System (Thermo Fisher Scientific; Massachusetts, United States)
5. Thermal cycler
6. Thunderbird SYBR qPCR Mix (Toyobo; Japan)
7. 96-well PCR plate (Roche Molecular Systems; California, United States)
8. Light Cycler 480 Real-Time PCR System (Roche Molecular Systems; California, United States)

2.3.3 Western blot

A. Preparation of lysate from cell culture

1. PBS
2. Lysis buffer (radioimmunoprecipitation (RIPA) assay buffer)
   - 150 mM sodium chloride (NaCl)
   - 1% nonyl phenoxypolyethoxylethanol (NP-40) or Triton X-100
   - 0.5% sodium deoxycholate
   - 0.1% sodium dodecyl sulfate (SDS)
   - 50 mM Tris, pH 8.0

B. Sample preparation

1. Coomassie protein assay reagent (Thermo Fisher Scientific; Massachusetts, United States)
2. 2 × Laemmli loading buffer:
   - 0.004% bromophenol blue
   - 10% 2-mercaptoethanol
   - 20% glycerol
   - 4% SDS
   - 0.125 M Tris(hydroxymethyl)aminomethane hydrochloride (Tris-HCL)

C. Sodium dodecyl-sulfate polyacrylamide gel electrophoresis (SDS-PAGE)

Note: The volumes mentioned are used for a single 8 × 13 cm gel.

1. Gel fix solution
   - 250 mL methanol
   - 50 mL glacial acetic acid
   - 200 mL water
2. Coomassie solution (filtered with Whatmann No. 1 filter paper)
   - 0.1% Coomassie Brilliant Blue R-250
   - 40% methanol
   - 10% glacial acetic acid
   - 500 mL water
3. De-stain solution
   - 250 mL methanol
   - 200 mL water
   - 50 mL glacial acetic acid

4. Gel storage solution
    - 25 mL glacial acetic acid
    - 475 mL water

D. Membrane transfer
1. Blotting paper
2. Polyvinylidene fluoride (PVDF) transfer membrane
3. Transfer buffer (25 mM Tris, 192 mM glycine, 20% methanol (v/v), pH 8.3)
4. 100% isopropanol
5. Tank transfer system
6. Roller

E. Immunodetection
1. Blocking buffer
    - 1 × TBST (1 × Tris-buffered saline (TBS) with 0.1% Tween 20 detergent)
    - 3%–5% nonfat dry milk or bovine serum albumin (BSA) in 1 × TBST
2. Primary antibody diluted with blocking buffer to working concentration
3. Secondary antibody diluted with blocking buffer to working concentration (conjugated with an enzyme such as horse radish peroxidase (HRP))
4. Enhanced chemiluminescent (ECL) substrate
5. TBST
6. Orbital shaker
7. Chemiluminescent imaging system (Biorad GelDoc Go Imaging System; California, United States).

3. Procedure

3.1 Cell culture

1. Place the vial containing the A549 cells in a water bath at 37°C for 30 s.
2. Mix 1.0 mL of cell suspension with complete growth media (at 37°C) in a 15 mL tube and centrifuge at 200 × *g* for 5 min using an Eppendorf Centrifuge 5804 (Mississauga, Canada).
3. After centrifugation, gently discard the supernatant.
4. Add 1.0 mL of fresh complete growth media to the tube and carefully resuspend the pellet.
5. Transfer the cells to a T-25 flask containing 5.0 mL of fresh complete growth media.
6. Examine the cells under an inverted microscope (Olympus CKX41; Germany).
7. Transfer the flask to a 5% CO_2 incubator at 37°C to allow the cell growth process.

3.2 Gene editing of stem cells

3.2.1 Cell seeding

1. Trypsinize the cells.

2. Perform cell count.
3. Seed cells at a density of 1×10^4 cells per well in a 96-well plate using a multichannel pipette.
4. Ensure uniform distribution of cells in the wells with each well containing a final volume of 1.0 mL.
5. Transfer the plate to an incubator with 5% CO_2 at 37°C for approximately 24 h until cells reached 70% confluency.

3.2.2 Transfection

1. Prepare two empty tubes for this protocol.
2. Prepare the first tube so that it contains a mixture of gRNA (50 ng), Opti-MEM medium (5.0 mL), TrueCut Cas9 protein v2 (250 ng), and Lipofectamine Cas9 Plus Reagent (0.5 mL).
3. Prepare a second tube by mixing Opti-MEM medium (5.0 mL) and Lipofectamine CRISPRMAX Cas9 Transfection Reagent (0.3 mL).
4. Incubate the second tube at room temperature (RT) for 1 min.
5. Gently pipette the content of the second tube into the first tube and mix it carefully.
6. Incubate the tube for 15 min at RT.
7. After incubation, gently add 10 μL of the mixture into each well of the 96-well plate containing cells.
8. Incubate the plate for 48 h in a CO_2 incubator at 37°C.
9. After incubation, gently remove the culture media and rinse cells with 50 μL of PBS.

3.3 Validation of CRISPR editing

3.3.1 T7 endonuclease 1 assay

A. T7E1 mismatch detection

1. Wash transfected cells with 100 μL PBS.
2. Add 50 μL of DNA extraction solution in each well of the 96-well plate to lyse cells.
3. Transfer cell lysate to PCR tubes.
4. Mix thoroughly and incubate at 65°C for 10 min, followed by 98°C for 5 min in a thermal cycler.
5. Add 100 μL of nuclease-free water to each of the PCR tubes.
6. Mix thoroughly and spin down.
7. Set up the PCR reactions containing DNA template, specific primers, components of the Alt-R Genome Editing Detection Kit, and PCR kit as follows:

Component	Amount: sample	Amount: Alt-R Control A	Amount: Alt-R Control B
Genomic DNA	4.0 μL (~40 ng)	–	–
Forward primer	300 nM	–	–
Reverse primer	300 nM	–	–
Alt-R Control A (template/primer mix)	–	2.0 μL	
Alt-R Control B (template/primer mix)	–	–	2.0 μL
PCR Buffer (5×)	5.0 μL (1×)	5.0 μL (1×)	5.0 μL (1×)
dNTPs	1.2 mM	1.2 mM	1.2 mM
PCR polymerase (1 U/μL)	0.5 U	0.5 U	0.5 U
Total volume	**25 μL**	**25 μL**	**25 μL**

8. Run the PCR according to the following setting:

Step	Temperature (°C)	Time (min:s)	Cycles
Predenaturation	95	5:00	1
Denaturation	95	0:20	30
Annealing	Primer specific[a]	0:15	
Extension	72	0:30	
Extension	72	2:00	1

[a]Annealing temperature for human HPRT primers is 67°C.

9. Form heteroduplexes for T7E1 digestion by combining the following in PCR tubes:

Component	Amount: sample	Amount: homoduplex control	Amount: heteroduplex control
PCR product	10 μL experimental target or Alt-R HPRT control	10 μL Control A	5.0 μL Control A 5.0 μL Control B
T7E1 Reaction buffer (10×)	2.0 μL	2.0 μL	2.0 μL
Nuclease-free water	6.0 μL	6.0 μL	6.0 μL
Total volume	**18 μL**	**18 μL**	**18 μL**

10. Incubate PCR products in a thermal cycler as follows:

Step	Temperature (°C)	Time
Denature	95	10 min
Ramp 1	95–85	Ramp rate −2°C/s
Ramp 2	85–25	Ramp rate −0.3°C/s

11. Perform T7E1 digestion in microcentrifuge tubes:

Component	Amount (μL)
PCR heteroduplexes	18
T7 endonuclease 1 (1 U/μL)	2.0
Total volume	**20**

12. Incubate the T7E1 reaction at 37°C for 60 min.
13. Visualize T7E1 mismatch detection results using agarose gel electrophoresis.

B. Agarose gel electrophoresis

1. Heat 0.4 g agarose gel powder in 40 mL 1 × Tris-Acetate buffer (TAE buffer) using a microwave for 1 min.
2. Left the mixture to cool to 60°C.
3. Add 1.0 μL ethidium bromide (EtBr) into the dissolved agarose mixture.
4. Pour the agarose mixture into a gel cast and leave it to harden for about 20–30 min at RT.
5. Combine 1 × loading dye with 1.0 μg genomic material and load it into gel.
6. Electrophorese at 90 V for 45 min.
7. Visualize the bands produced under ultraviolet (UV) light (Fig. 4).

3.3.2 RT-qPCR

A. RNA extraction

1. Discard culture medium from cell monolayer.
2. Trypsinize cells by incubating cells with 0.25% trypsin-EDTA, followed by centrifugation at 300 × *g* for 5 min.
3. Add 350 μL lysis buffer to pelleted cells (lesser than 5×10^6 cells).
4. Alternatively, for direct lysis of cells, add 350 μL lysis buffer directly onto the cell monolayer.
5. Collect and transfer cell lysate into a microcentrifuge tube.
6. Vortex to mix cell lysate thoroughly.
7. Add 350 μL of 70% ethanol to cell lysate and mix thoroughly.
8. Transfer the content into the spin column placed in a 2.0 mL collection tube.
9. Close the lid gently and centrifuge the tube at ≥8000 × *g* for 15 s. Discard the flow-through after centrifugation.

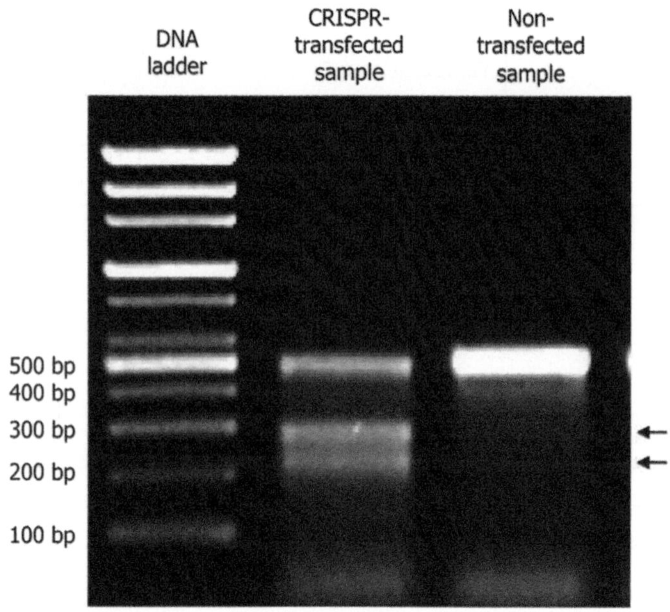

FIG. 4

T7E1 mismatch detection results. CRISPR-transfected cells and nontransfected cells were treated with T7E1 endonuclease, and the products were analyzed on a 2% agarose gel. Nontransfected sample showed one band at 500 bp. CRISPR-transfected sample also showed bands of ~280 bp and ~220 bp resulting from T7E1 endonuclease cleavage *(arrows)*.

Adapted and modified from Kim, E.K., Kim, S., Maeng, Y.S., 2019. Generation of TGFBI knockout ABCG2+/ABCB5+ double-positive limbal epithelial stem cells by CRISPR/Cas9-mediated genome editing. PLoS One 14(2), e0211864.

10. Add 700 μL buffer to the RNeasy spin column and centrifuge the tube for 15 s at ≥8000 × *g*. Discard the flow-through.
11. Add 500 μL of washing buffer to the spin column and centrifuge the tube at ≥8000 × *g* for 2 min. Discard the flow-through. Repeat this step twice. This repeated centrifugation step is to ensure the spin column membrane is dried enough so that none of the ethanol is carried over. Beware that ethanol contamination may interfere with downstream reactions.
12. Transfer the spin column into a fresh 1.5 mL microcentrifuge tube. Add 30 μL RNase-free water onto the column membrane.
13. Centrifuge the collection tube for 1 min at 8000 × *g* to elute the RNA.
14. Determine the concentration and quality of yielded RNA using Nanodrop and gel electrophoresis. Pure RNA should show absorbance value (A_{260}/A_{280}) of 1.8–2.0.

B. Complementary DNA (cDNA) synthesis

1. Combine the extracted RNA and 2.0 μM primers (1.0 μM forward primer and 1.0 μM reverse primer) in a 1.5 mL microcentrifuge tube and briefly centrifuge the tube.
2. Heat the tube in a thermal cycler at 65°C for 5 min and chill it on ice for 2 min.
3. Combine 13 μL annealed RNA with 5.0 μL reverse transcription mix.
4. Incubate the tube at 55°C for 10 min and 80°C for 10 min.
5. Store the reaction mixture at −20°C or proceed with RT-qPCR immediately.

C. Quantitative polymerase chain reaction (qPCR)

1. Mix cDNA (80 ng), qPCR mix (4.0 μL), and primers (0.3 μM forward primer and 0.3 μM reverse primer) in a 96-well PCR plate. Top up the reaction mixture to 20 μL with water for each well.
2. Briefly spin down the reaction mix before placing the plate onto the Light Cycler 480 Real-Time PCR System.
3. Carry out the amplification of cDNA according to the following setting:

Step	Temperature (°C)	Time (min:s)	Cycles
Predenaturation	95	1:00	1
Denaturation	95	0:15	40
Annealing	Primer-specific	0:30	
Extension	72	1:00	

4. Key in the sample information including control wells in the Light Cycler 480 software (Version 1.5) and press "Run" to start the cDNA amplification.
5. After the run is completed, calculate relative fold change in gene expression using $2^{-\Delta\Delta Ct}$ method, where $\Delta\Delta CT = \Delta CT$(a target sample) − ΔCT(a reference sample) and $\Delta Ct = Ct$(gene of interest) − Ct(housekeeping gene).

3.3.3 Western blot

A. Preparation of lysate from cell culture

1. Discard culture medium and wash the cells with ice-cold PBS.
2. Discard the PBS, add ice-cold lysis buffer (1.0 mL per 10^7 cells or 100 mm dish or 150 cm^2 flask).
3. Transfer the cell lysate into a 1.5 mL microcentrifuge tube.
4. Alternatively, trypsinize the cells and wash the pelleted cells with ice-cold PBS.
5. Suspend the cells into lysis buffer and agitate the microcentrifuge tubes for 30 min at 4°C.
6. Centrifuge the tubes at 16,000 × *g* for 20 min at 4°C.
7. Collect the supernatant in fresh tube and place on ice. Discard the pellet.

B. Sample preparation

1. Perform a protein quantification assay with a small volume of lysate using Coomassie protein assay reagent by comparing absorbance reading at 280 nm with the standards.
2. Determine loading protein concentration and add an equal volume of 2× Laemmli loading buffer.
3. Boil the cell lysate at 95°C for 5 min. Centrifuge at 16,000 × *g* for 5 min.
4. Store the samples at −20°C or proceed with gel electrophoresis immediately.

C. SDS-PAGE

1. Prepare resolving gel solution using the following volumes (for 10 mL) depending on the size of protein of interest:

Protein size (kDa)	Gel (%)	Water (mL)	30% acrylamide (mL)	1.5 M Tris-HCl, pH 8.8 (mL)	10% SDS (μL)	10% ammonium persulfate (APS) (μL)	Tetra methylethy lenediamine (TEMED)[a] (μL)
25–100	8	4.6	2.6	2.6	100	100	10
15–100	10	3.8	3.4	2.6	100	100	10
12–45	15	2.2	5.0	2.6	100	100	10

[a]*TEMED must be added last.*

2. Allow the gel to set for 30 min at RT.
3. Pour the gel solution in the plates assembled with spacers. Cover the surface with water or isopropanol to maintain an even gel surface.
4. Allow the gel to set for about 20–30 min at RT.
5. Discard the overlayed water or isopropanol on the resolving gel.
6. Prepare stacking gel solution using the following volumes (for 10 mL):

Gel (%)	Water (mL)	30% acrylamide (mL)	0.5 M Tris-HCl, pH 6.8 (mL)	10% SDS (μL)	10% APS (μL)	TEMED[a] (μL)
5	5.86	1.34	2.6	100	100	10

[a]*TEMED must be added last*

7. Add the stacking gel solution until it overflows. Insert the comb and ensure no formation of air bubbles.
8. Allow the gel to set for 30 min at RT.
9. Load equal amounts of protein (20–30 μg) into the wells.
10. Run the gel for 1–2 h at 100 V.
11. After electrophoresis, transfer the gel into gel fix solution and fix the proteins on rocking shaker for 2 h.
12. Discard the gel fix solution and replace with Coomassie solution. Stain the gel on rocking shaker for 2–4 h.

13. Remove excess stain by washing the gel twice with distilled water.
14. Add de-stain solution to the gel and incubate on a rocking shaker for another 4h until visible blue bands appear.
15. The gel may be stored in gel storage solution.

D. Membrane transfer

1. Prepare the PVDF membrane by wetting in methanol for 15 s.
2. Assemble the stack in cassette holder as illustrated in Fig. 5.
3. Blot roller can be used to ensure even surface of each layer in the stack.
4. Run the transfer tank system for 1 to 2 h at 6 to 8 V/cm interelectrode distance.
5. Remove the cassette holder after complete transfer and disassemble the transfer stack using forceps.

E. Immunodetection

1. Block the membrane for 1 h at RT using blocking buffer.
2. Incubate the membrane with primary antibody for 1 h at RT.
3. Wash the membrane in TBST thrice, with 5 min each wash.
4. Incubate the membrane with secondary antibody for 1 h at RT.
5. Wash the membrane in TBST thrice, with 5 min each wash.
6. Incubate the membrane in the ECL substrate for 1–5 min.
7. Remove the membrane from the ECL substrate and sandwich it between layers of plastic sheet protector, ensuring no formation of bubbles in between the layers.
8. Place a sheet of autoradiography film on top and place the wrapped membrane in a film cassette.
9. In dark, image the membrane with a chemiluminescent imaging system.

FIG. 5

Assembly of SDS-PAGE stack. Place the cassette holder horizontally to perform the stacking. First, place sponge into the cassette holder beside the negative electrode and lay filter paper on top. Then, place the gel carefully and followed by PVDF membrane. Place filter paper on the stack and finish the assembly with sponge. Close the cassette holder and return it to its vertical orientation.

4. Analyzation after gene editing: Transcriptomic analysis

4.1 RNA-Seq and data processing pipelines

A typical NGS pipeline or workflow includes aligning and assembling the sequenced reads to a reference, quantifying the transcripts, normalizing the expression values, and then analyzing the data downstream (Fig. 6).

Once a list of genes and their associated expression values has been generated, researchers will then be able to utilize yet another array of bioinformatics tools in order to perform gene expression profiling and make meaningful interpretations of the result. Overrepresentation analysis (ORA) is one of the most convenient methods that belongs to the first-generation of RNA-Seq tools for identifying the most affected pathways (Khatri et al., 2012). It works on the assumption that pathways with a higher number of differentially expressed genes (DEGs) are more likely to be statistically significant (García-Campos et al., 2015). The data can then be

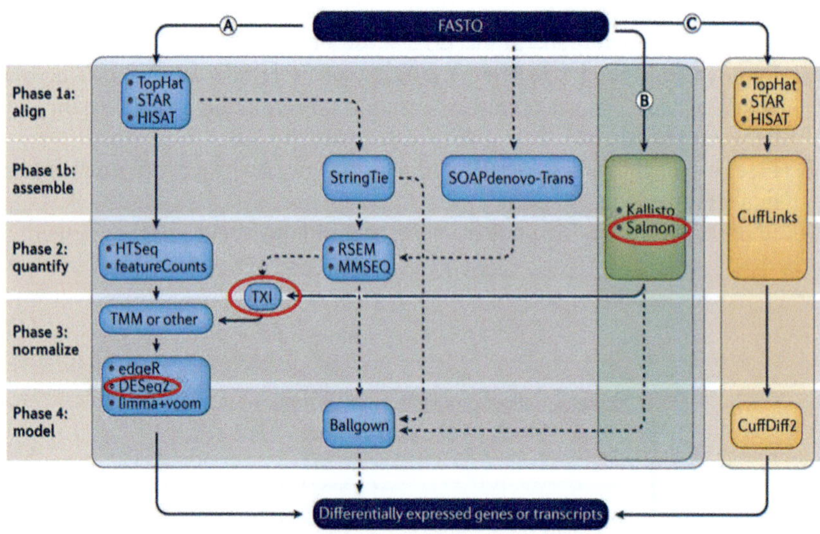

FIG. 6

A typical NGS pipeline used in RNA-Seq. (A) A common pipeline that uses a reference genome to quantify and identify gene locations. (B) A newer pipeline utilizing transcriptome mapping to conveniently quantify transcript abundance (*red indicators* show common Salmon tool pipelines). (C) Another variation of data analysis that uses a streamlined set of tools for RNA-Seq analysis. Finally, the list of genes can then be analyzed further using downstream analytical tools.

Adapted from Stark, R., Grzelak, M., Hadfield, J., 2019. RNA sequencing: the teenage years. Nat. Rev. Genet. 20(11), 631–656.

visualized using a wide variety of mediums such as scatter plots and Venn diagrams. Most ORA tools can be performed online, such as on the Gene Ontology website (Blake et al., 2015). However, the main limitation of ORA is that every gene is analyzed independently regardless of gene interactions, and it relies on user-defined cutoff values (usually $P < .05$ and fold change $> \pm 2.0$). In order to improve on this, a second-generation method known as functional enrichment analysis was devised. This utilized a functional class scoring (FCS) method that could detect enriched pathways using all the quantitative data from the list of DEGs without the need for user-defined cutoff values (Zyla et al., 2017). This would result in increased sensitivity for detecting significant pathways that would otherwise be discarded when a researcher uses cutoff values that are too stringent. However, this method does not convey directional information of the pathway (i.e., activated or inactivated). Undoubtedly, one of the most popular tools from this category is the Gene Set Enrichment Analysis tool (Zyla et al., 2017).

Pathway topology analysis is the most recent generation of RNA-Seq analysis tool. It analyzes a gene's position on a network plus its quantitative data to determine if any associated pathways are significant (Han et al., 2015). This is because a pathway consists of not only activators but also effectors, repressors, and co-regulating factors from other pathways. As such, these should be considered in order to determine statistical significance of a particular pathway. Kyoto Encyclopedia of Genes and Genomes (KEGG) is one such database that contains important pathway information as a reference (Kanehisa et al., 2017). There are several tools under this category that emphasize the different kinds of gene pathway members during analysis (Fig. 7). The signaling pathway impact analysis (SPIA) tool is one such example that focuses on genes with many regulatory interactions such as ligands and receptors (e.g., transforming growth factor alpha [TGFA]) (Ihnatova et al., 2018). Currently, second-generation FCS methods and third-generation pathway topology methods are both highly effective but the latter performs better with nonoverlapping pathways containing unique genes (Bayerlová et al., 2015). Ultimately, no one method trumps the other, and continuous development is constantly improving the accuracy and sensitivity of these tools.

5. Conclusion

CRISPR-Cas9 is an efficient tool for use in modifying gene expression in various types of cells, including stem cells. The proper selection of gRNA will reduce off-target mutagenesis, and with well-optimized transfection protocol, the efficiency of gene modification will be significantly enhanced. Transcriptomic analysis from genetically edited cells would be useful to identify the direct effect of the modified gene on the downstream signaling pathways, hence revealing possible novel connection of a gene to a cellular function.

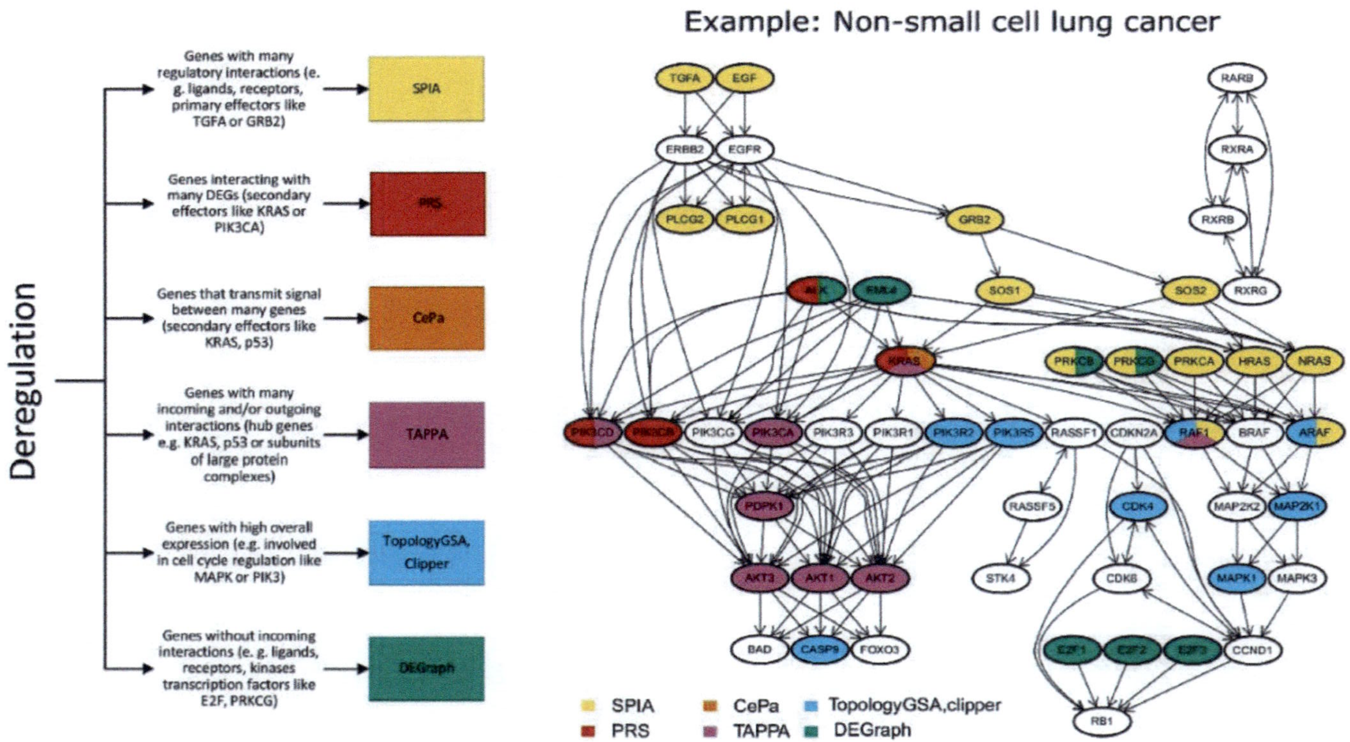

FIG. 7

Examples of pathway topology tools used based on different aspects of a pathway. Depending on the focus of the study, deregulated genes of interest can be analyzed based on the different tools currently available to the researchers. Optimizing the method for analyzing pathways can improve the performance of a particular tool as these methods heavily depend on gene topological properties.

Adapted from Bayerlová, M., Jung, K., Kramer, F., Klemm, F., Bleckmann, A., Beißbarth, T., 2015. Comparative study on gene set and pathway topology-based enrichment methods. BMC Bioinformatics 16, 334.

Abbreviations

APS	ammonium persulfate
BSA	bovine serum albumin
Cas	clustered regularly interspaced short palindromic repeats associated
cDNA	complementary DNA
CRISPR	clustered regularly interspaced short palindromic repeats
crRNA	CRISPR RNA
DEG	differentially expressed gene
DNA	deoxyribonucleic acid
DSB	double-strand break
ECL	enhanced chemiluminescent
EDTA	ethylenediaminetetraacetic acid
EtBr	ethidium bromide
FBS	fetal bovine serum
FCS	functional class scoring
HDR	homology-directed repair
HPRT	hypoxanthine phosphoribosyl transferase
HRP	horseradish peroxidase
HSC	hematopoietic stem cell
KEGG	Kyoto Encyclopedia of Genes and Genomes
NaCl	sodium chloride
NGS	next-generation sequencing
NHEJ	nonhomologous end joining
NP-40	nonyl phenoxypolyethoxylethanol
Opti-MEM	Opti-Minimal Essential Medium
ORA	over-representation analysis
PAM	protospacer adjacent motif
PBS	phosphate-buffered saline
PCR	polymerase chain reaction
PS	penicillin–streptomycin
PVDF	polyvinylidene fluoride
RIPA	radioimmunoprecipitation
RNA	ribonucleic acid
RPMI	Roswell Park Memorial Institute
RT	room temperature
RT-qPCR	reverse transcription quantitative polymerase chain reaction
SDS	sodium dodecyl sulfate
SDS-PAGE	Sodium dodecyl-sulfate polyacrylamide gel electrophoresis
sgRNA	single guide RNA
SPIA	Signaling Pathway Impact Analysis
TAE buffer	tris-acetate buffer
TBS	tris-buffered saline
TEMED	tetramethylethylenediamine
TGFA	transforming growth factor alpha
tracrRNA	trans-activating RNA
Tris-HCL	tris(hydroxymethyl)aminomethane hydrochloride
T7E1	T7 endonuclease 1
UV	ultraviolet

References

Bayerlová, M., Jung, K., Kramer, F., Klemm, F., Bleckmann, A., Beißbarth, T., 2015. Comparative study on gene set and pathway topology-based enrichment methods. BMC Bioinformatics 16, 334.

Bennardo, N., Cheng, A., Huang, N., Stark, J.M., 2008. Alternative-NHEJ is a mechanistically distinct pathway of mammalian chromosome break repair. PLoS Genet. 4, e1000110.

Blake, J.A., Christie, K.R., Dolan, M.E., Drabkin, H.J., Hill, D.P., Ni, L., Sitnikov, D., Burgess, S., Buza, T., Gresham, C., McCarthy, F., Pillai, L., Wang, H., Carbon, S., Dietze, H., Lewis, S.E., Mungall, C.J., Munoz-Torres, M.C., Feuermann, M., Gaudet, P., Basu, S., Chisholm, R.L., Dodson, R.J., Fey, P., Mi, H., Thomas, P.D., Muruganujan, A., Poudel, S., Hu, J.C., Aleksander, S.A., McIntosh, B.K., Renfro, D.P., Siegele, D.A., Attrill, H., Brown, N.H., Tweedie, S., Lomax, J., Osumi-Sutherland, D., Parkinson, H., Roncaglia, P., Lovering, R.C., Talmud, P.J., Humphries, S.E., Denny, P., Campbell, N.H., Foulger, R.E., Chibucos, M.C., Giglio, M.G., Chang, H.Y., Finn, R., Fraser, M., Mitchell, A., Nuka, G., Pesseat, S., Sangrador, A., Scheremetjew, M., Young, S.Y., Stephan, R., Harris, M.A., Oliver, S.G., Rutherford, K., Wood, V., Bahler, J., Lock, A., Kersey, P.J., McDowall, M. D., Staines, D.M., Dwinell, M., Shimoyama, M., Laulederkind, S., Hayman, G.T., Wang, S.J., Petri, V., D'Eustachio, P., Matthews, L., Balakrishnan, R., Binkley, G., Cherry, J.M., Costanzo, M.C., Demeter, J., Dwight, S.S., Engel, S.R., Hitz, B.C., Inglis, D.O., Lloyd, P., Miyasato, S.R., Paskov, K., Roe, G., Simison, M., Nash, R.S., Skrzypek, M.S., Weng, S., Wong, E.D., Berardini, T.Z., Li, D., Huala, E., Argasinska, J., Arighi, C., Auchincloss, A., Axelsen, K., Argoud-Puy, G., Bateman, A., Bely, B., Blatter, M.C., Bonilla, C., Bougueleret, L., Boutet, E., Breuza, L., Bridge, A., Britto, R., Casals, C., Cibrian-Uhalte, E., Coudert, E., Cusin, I., Duek-Roggli, P., Estreicher, A., Famiglietti, L., Gane, P., Garmiri, P., Gos, A., Gruaz-Gumowski, N., Hatton-Ellis, E., Hinz, U., Hulo, C., Huntley, R., Jungo, F., Keller, G., Laiho, K., Lemercier, P., Lieberherr, D., Macdougall, A., Magrane, M., Martin, M., Masson, P., Mutowo, P., O'Donovan, C., Pedruzzi, I., Pichler, K., Poggioli, D., Poux, S., Rivoire, C., Roechert, B., Sawford, T., Schneider, M., Shypitsyna, A., Stutz, A., Sundaram, S., Tognolli, M., Wu, C., Xenarios, I., Chan, J., Kishore, R., Sternberg, P.W., Van Auken, K., Muller, H.M., Done, J., Li, Y., Howe, D., Westerfeld, M., 2015. Gene ontology consortium: going forward. Nucleic Acids Res. 43, D1049–D1056.

Brunger, J.M., Zutshi, A., Willard, V.P., Gersbach, C.A., Guilak, F., 2017. Genome engineering of stem cells for autonomously regulated, closed-loop delivery of biologic drugs. Stem Cell Rep. 8, 1202–1213.

Cong, L., Ran, F.A., Cox, D., Lin, S., Barretto, R., Habib, N., Hsu, P.D., Wu, X., Jiang, W., Marraffini, L.A., Zhang, F., 2013. Multiplex genome engineering using CRISPR/Cas systems. Science 339, 819–823.

Dever, D.P., Bak, R.O., Reinisch, A., Camarena, J., Washington, G., Nicolas, C.E., Pavel-Dinu, M., Saxena, N., Wilkens, A.B., Mantri, S., Uchida, N., Hendel, A., Narla, A., Majeti, R., Weinberg, K.I., Porteus, M.H., 2016. CRISPR/Cas9 β-globin gene targeting in human haematopoietic stem cells. Nature 539, 384–389.

Frangoul, H., Altshuler, D., Cappellini, M.D., Chen, Y.S., Domm, J., Eustace, B.K., Foell, J., de la Fuente, J., Grupp, S., Handgretinger, R., Ho, T.W., Kattamis, A., Kernytsky, A., Lekstrom-Himes, J., Li, A.M., Locatelli, F., Mapara, M.Y., de Montalembert, M., Rondelli, D., Sharma, A., Sheth, S., Soni, S., Steinberg, M.H., Wall, D., Yen, A., Corbacioglu, S., 2021. CRISPR-Cas9 gene editing for sickle cell disease and β-thalassemia. N. Engl. J. Med. 384 (3), 252–260. https:/doi.org/10.1056/NEJMoa2031054.

García-Campos, M.A., Espinal-Enríquez, J., Hernández-Lemus, E., 2015. Pathway analysis: state of the art. Front. Physiol. 6, 383.

Han, Y., Gao, S., Muegge, K., Zhang, W., Zhou, B., 2015. Advanced applications of RNA sequencing and challenges. Bioinform. Biol. Insights 9, 29–46.

Han, W., She, Q., 2017. CRISPR history: discovery, characterization, and prosperity. In: Progress in Molecular Biology and Translational Science. Elsevier B.V, pp. 1–21.

Hoellerbauer, P.F., Paddison, P.J., Cooper, J.A., Holland, E.C., 2020. Identification of Genetic Vulnerabilities in Glioblastoma and Other Cancers Using CRISPR-Cas9-Based Functional Genomic Approaches.

Horvath, P., Romero, D.A., Coûté-Monvoisin, A.C., Richards, M., Deveau, H., Moineau, S., Boyaval, P., Fremaux, C., Barrangou, R., 2008. Diversity, activity, and evolution of CRISPR loci in Streptococcus thermophilus. J. Bacteriol. 190, 1401–1412.

Hsu, P.D., Lander, E.S., Zhang, F., 2014. Development and applications of CRISPR-Cas9 for genome engineering. Cell 157, 1262–1278.

Ihnatova, I., Popovici, V., Budinska, E., 2018. A critical comparison of topology-based pathway analysis methods. PLoS One 13, e0191154.

Ishino, Y., Shinagawa, H., Makino, K., Amemura, M., Nakatura, A., 1987. Nucleotide sequence of the iap gene, responsible for alkaline phosphatase isoenzyme conversion in Escherichia coli, and identification of the gene product. J. Bacteriol. 169, 5429–5433.

Jansen, R., Van Embden, J.D.A., Gaastra, W., Schouls, L.M., 2002. Identification of genes that are associated with DNA repeats in prokaryotes. Mol. Microbiol. 43, 1565–1575.

Jiang, F., Liang, Y., Wei, W., Zou, C., Chen, G., Wan, Y., Liu, Z., Yang, Y., Han, Z., Zhu, J., Zhong, W., 2020. Functional classification of prostate cancer-associated miRNAs through CRISPR/Cas9-mediated gene knockout. Mol. Med. Rep. 22 (5), 3777–3784.

Kanehisa, M., Furumichi, M., Tanabe, M., Sato, Y., Morishima, K., 2017. KEGG: new perspectives on genomes, pathways, diseases and drugs. Nucleic Acids Res. 45, D353–D361.

Karvelis, T., Gasiunas, G., Miksys, A., Barrangou, R., Horvath, P., Siksnys, V., 2013. crRNA and tracrRNA guide Cas9-mediated DNA interference in *Streptococcus thermophilus*. RNA Biol. 10, 841–851.

Kennedy, E.M., Kornepati, A.V.R., Goldstein, M., Bogerd, H.P., Poling, B.C., Whisnant, A. W., Kastan, M.B., Cullen, B.R., 2014. Inactivation of the human papillomavirus E6 or E7 gene in cervical carcinoma cells by using a bacterial CRISPR/Cas RNA-guided endonuclease. J. Virol. 88, 11965–11972.

Khatri, P., Sirota, M., Butte, A.J., 2012. Ten years of pathway analysis: current approaches and outstanding challenges. PLoS Comput. Biol. 8, e1002375.

Kistler, K.E., Vosshall, L.B., Matthews, B.J., 2015. Genome engineering with CRISPR-Cas9 in the mosquito aedes aegypti. Cell Rep. 11, 51–60.

Le Rhun, A., Escalera-Maurer, A., Bratovič, M., Charpentier, E., 2019. CRISPR-Cas in streptococcus pyogenes. RNA Biol. 16, 380–389.

Lee, J.K., Jeong, E., Lee, J., Jung, M., Shin, E., Kim, Y.H., Lee, K., Jung, I., Kim, D., Kim, S., Kim, J.S., 2018. Directed evolution of CRISPR-Cas9 to increase its specificity. Nat. Commun. 9, 1–10.

Makarova, K.S., Haft, D.H., Barrangou, R., Brouns, S.J.J., Charpentier, E., Horvath, P., Moineau, S., Mojica, F.J.M., Wolf, Y.I., Yakunin, A.F., Van Der Oost, J., Koonin, E.V., 2011. Evolution and classification of the CRISPR-Cas systems. Nat. Rev. Microbiol. 9, 467–477.

Miyaoka, Y., Berman, J.R., Cooper, S.B., Mayerl, S.J., Chan, A.H., Zhang, B., Karlin-Neumann, G.A., Conklin, B.R., 2016. Systematic quantification of HDR and NHEJ reveals effects of locus, nuclease, and cell type on genome-editing. Sci. Rep. 6, 1–12.

Mojica, F.J.M., Juez, G., Rodriguez-Valera, F., 1993. Transcription at different salinities of Haloferax mediterranei sequences adjacent to partially modified PstI sites. Mol. Microbiol. 9, 613–621.

Mojica, F.J.M., Díez-Villaseñor, C., García-Martínez, J., Soria, E., 2005. Intervening sequences of regularly spaced prokaryotic repeats derive from foreign genetic elements. J. Mol. Evol. 60, 174–182.

Ou, Z., Niu, X., He, W., Chen, Y., Song, B., Xian, Y., Fan, D., Tang, D., Sun, X., 2016. The combination of CRISPR/Cas9 and iPSC technologies in the gene therapy of human β-thalassemia in mice. Sci. Rep. 6, 32463.

Pankowicz, F.P., Barzi, M., Legras, X., Hubert, L., Mi, T., Tomolonis, J.A., Ravishankar, M., Sun, Q., Yang, D., Borowiak, M., Sumazin, P., Elsea, S.H., Bissig-Choisat, B., Bissig, K. D., 2016. Reprogramming metabolic pathways in vivo with CRISPR/Cas9 genome editing to treat hereditary tyrosinaemia. Nat. Commun. 7, 12642.

Puschnik, A.S., Majzoub, K., Ooi, Y.S., Carette, J.E., 2017. A CRISPR toolbox to study virus-host interactions. Nat. Rev. Microbiol. 15 (6), 351–364.

Rath, D., Amlinger, L., Rath, A., Lundgren, M., 2015. The CRISPR-Cas immune system: biology, mechanisms and applications. Biochimie 117, 119–128.

Schwank, G., Koo, B.K., Sasselli, V., Dekkers, J.F., Heo, I., Demircan, T., Sasaki, N., Boymans, S., Cuppen, E., Van Der Ent, C.K., Nieuwenhuis, E.E.S., Beekman, J.M., Clevers, H., 2013. Functional repair of CFTR by CRISPR/Cas9 in intestinal stem cell organoids of cystic fibrosis patients. Cell Stem Cell 13, 653–658.

Shmakov, S., Abudayyeh, O.O., Makarova, K.S., Wolf, Y.I., Gootenberg, J.S., Semenova, E., Minakhin, L., Joung, J., Konermann, S., Severinov, K., Zhang, F., Koonin, E.V., 2015. Discovery and functional characterization of diverse class 2 CRISPR-Cas systems. Mol. Cell 60, 385–397.

Tang, S., Chen, T., Yu, Z., Zhu, X., Yang, M., Xie, B., Li, N., Cao, X., Wang, J., 2014. RasGRP3 limits toll-like receptor-triggered inflammatory response in macrophages by activating Rap1 small GTPase. Nat. Commun. 5, 4657.

Vouillot, L., Thélie, A., Pollet, N., 2015. Comparison of T7E1 and surveyor mismatch cleavage assays to detect mutations triggered by engineered nucleases. G3: genes, genomes. Genetics 5 (3), 407–415.

Wu, X., Scott, D.A., Kriz, A.J., Chiu, A.C., Hsu, P.D., Dadon, D.B., Cheng, A.W., Trevino, A. E., Konermann, S., Chen, S., Jaenisch, R., Zhang, F., Sharp, P.A., 2014. Genome-wide binding of the CRISPR endonuclease Cas9 in mammalian cells. Nat. Biotechnol. 32, 670–676.

Yin, H., Xue, W., Chen, S., Bogorad, R.L., Benedetti, E., Grompe, M., Koteliansky, V., Sharp, P.A., Jacks, T., Anderson, D.G., 2014. Genome editing with Cas9 in adult mice corrects a disease mutation and phenotype. Nat. Biotechnol. 32, 551–553.

Zhang, C., Xiao, B., Jiang, Y., Zhao, Y., Li, Z., Gao, H., Ling, Y., Wei, J., Li, S., Lu, M., Su, X. Z., Cui, H., Yuan, J., 2014. Efficient editing of malaria parasite genome using the CRISPR/Cas9 system. MBio 5, e01414-14.

Zhang, N., Zhi, H., Curtis, B.R., Rao, S., Jobaliya, C., Poncz, M., French, D.L., Newman, P.J., 2016. CRISPR/Cas9-mediated conversion of human platelet alloantigen allotypes. Blood 127, 675–680.

Zhen, S., Hua, L., Takahashi, Y., Narita, S., Liu, Y.H., Li, Y., 2014. In vitro and in vivo growth suppression of human papillomavirus 16-positive cervical cancer cells by CRISPR/Cas9. Biochem. Biophys. Res. Commun. 450, 1422–1426.

Zuo, E., Cai, Y.J., Li, K., Wei, Y., Wang, B.A., Sun, Y., Liu, Z., Liu, J., Hu, X., Wei, W., Huo, X., Shi, L., Tang, C., Liang, D., Wang, Y., Nie, Y.H., Zhang, C.C., Yao, X., Wang, X., Zhou, C., Ying, W., Wang, Q., Chen, R.C., Shen, Q., Xu, G.L., Li, J., Sun, Q., Xiong, Z.Q., Yang, H., 2017. One-step generation of complete gene knockout mice and monkeys by CRISPR/Cas9-mediated gene editing with multiple sgRNAs. Cell Res. 27, 933–945.

Zyla, J., Marczyk, M., Weiner, J., Polanska, J., 2017. Ranking metrics in gene set enrichment analysis: do they matter? BMC Bioinformatics 18, 256.

Further reading

Chandrasekaran, A.P., Song, M., Kim, K.S., Ramakrishna, S., 2018. Different methods of delivering CRISPR/Cas9 into cells. In: Progress in Molecular Biology and Translational Science. Elsevier B.V, pp. 157–176.

Jones, M.K., Lu, B., Saghizadeh, M., Wang, S., 2016. Gene expression changes in the retina following subretinal injection of human neural progenitor cells into a rodent model for retinal degeneration. Mol. Vis. 22, 472–490.

Pisarska, M.D., Akhlaghpour, M., Lee, B., Barlow, G.M., Xu, N., Wang, E.T., Mackey, A.J., Farber, C.R., Rich, S.S., Rotter, J.I., Chen, Y.D.I., Goodarzi, M.O., Guller, S., Williams, J., 2016. Optimization of techniques for multiple platform testing in small, precious samples such as human chorionic villus sampling. Prenat. Diagn. 36, 1061–1070.

Son, K., Yu, S., Shin, W., Han, K., Kang, K., 2018. A simple guideline to assess the characteristics of RNA-Seq data. Biomed. Res. Int. 2018, 2906292.

Sürün, D., von Melchner, H., Schnütgen, F., 2018. CRISPR/Cas9 genome engineering in hematopoietic cells. Drug Discov. Today Technol. 28, 33–39.

Wang, J., Geisert, E.E., Struebing, F.L., 2019. RNA sequencing profiling of the retina in C57BL/6J and DBA/2J mice: enhancing the retinal microarray data sets from GeneNetwork. Mol. Vis. 25, 345–358.

CHAPTER

Hypoxic environment: A new solution to stem cell culture

9

Ghee Chien Ooi[a], Pooi Ling Mok[b], and Suresh Kumar Subbiah[c]

[a]*CryoCord Sdn. Bhd., Cyberjaya, Selangor, Malaysia,* [b]*Department of Biomedical Science, Faculty of Medicine and Health Sciences, Universiti Putra Malaysia, Serdang, Selangor, Malaysia,* [c]*Centre for Materials Engineering and Regenerative Medicine, Bharath Institute of Higher Education and Research, Chennai, Tamil Nadu, India*

1. Introduction

Stem cells are cells that have the potential to develop into different types of cells. The unique characteristic of a stem cell is that they have the ability to self-renew and can differentiate into different types of tissue or organ with specific functions. Due to its potential, it has emerged as one of the promising sources for tissue engineering and regenerative medicine.

The previous chapters have identified various methods of extraction or isolation of stem cells from different tissues. The percentage of isolated cells is low, thus requiring us to massively expand the culture in the laboratory for downstream application including clinical transplantation. During this procedure, cells can attain different attributes when they are exposed to a new physical and biological microenvironment deviating from its origin. An incubator is used to provide a suitable physical environment, particularly temperature, composition of air, and humidity, for cell growth.

Diffusion of oxygen gas to the deep and distant tissues in our body will be reduced when blood is transported from the heart. The partial pressure of oxygen in different tissues depends on its physiological and pathological states, and physical-chemical processes in our body. For example, the brain, bone marrow, and umbilical arteries have 30–48 mmHg, 51.8 ± 14.5 mmHg, and 18 mmHg, respectively (Ortiz-Prado et al., 2019). The stem cells isolated from various tissues in human and canine demonstrated significant gene expression profiles and affected the proliferative and differentiation potential, and probably the functions as well (Wagner et al., 2005). The blastocyst in which the inner cell mass is harvested to form embryonic stem cells resides in hypoxic condition during human development until gestational week of 12 to 13 (Haque et al., 2013). Hence, using an ambient 21% oxygen gas supply for maintaining a cell culture may not set an appropriate condition for the stem cells to grow without changing their intrinsic properties. The emergence

Stem Cell Laboratory Techniques. https://doi.org/10.1016/B978-0-12-823729-8.00001-4

of hypoxic incubator has flouted the "one size fits all" rule that we should use the same oxygen pressure setting for all types of stem cell cultures in our practice, ignoring the fact that they are derived from various tissue.

It should be noted that in the end a stem cell expansion aims to deliver sufficient number of cells to effect a desirable clinical primary outcome. In most cases, the microenvironment sustaining injury is pathologically harsh. This poses a challenge to clinicians as the delivered stem cells may not be able to survive in this condition. By priming the stem cells in hypoxic culture condition, the cells will better arm themselves with higher expression of pro-survival genes such as Akt, HO-1, and Hsp70, hence improving the result outcomes (Tsai et al., 2012). Therefore, this chapter will discuss the influence of the hypoxic culture environment on the cells, in addition to providing insights into the selection, operation, and maintenance of a hypoxic incubator for new users.

2. Influence of hypoxia condition on stem cell culture

A hypoxic environment is defined as a surrounding condition that has a low oxygen concentration. The aim of culturing stem cell under hypoxic environment is to mimic the in vitro physiological conditions. Since mesenchymal stem cells (MSCs) can be isolated from various sources such as dental pulp, umbilical cord, bone marrow, and adipose tissue with different niche environments, it is important to study at which concentration of O_2 will allow us to harness its best potential for clinical application.

Based on a study reported by Ahmed et al., the dental pulp stem cells exhibited relatively smaller, spindle-shaped cells when cultured in hypoxic condition when compared to the normoxic environment, which showed larger and flattened morphology. In addition, stem cell culture in hypoxic environment showed higher expression of stem cell markers, enhanced cell proliferation, improved migration capability, and trophic effect of the secretome (Ahmed et al., 2016).

Provision of a hypoxic environment has also been shown to increase proliferation capacity, ameliorate cell damage due to the reduction of reactive oxidative stress (ROS) and increase in glucose consumption and lactate production in umbilical cord-derived human MSCs (Lavrentieva et al., 2010). The benefits of hypoxic environment for stem cell culture do not limit to umbilical cord-derived cells only. It also showed that Wharton Jelly-derived MSCs could be coaxed to secrete more cytokines or trophic factors when incubated in a hypoxic environment (Teixeira et al., 2015).

In addition, MSCs harvested from the bone marrow demonstrated short-term proliferation, long-term expansion efficiency, and maintenance of cell stemness (Hung, 2016). In another article, supplying low oxygen gas to the bone marrow MSCs culture can help to modulate paracrine secretion capacity and upregulate the expression of chemokine receptors CXCR4, CXCR7, and CX3CR1 (Haque et al., 2013). These receptors will bind to the chemokines released from the injured tissue, thereby helping in cell migration and engraftment ability (Hung, 2013).

As for human adipose-derived stem cells, hypoxic culture in the presence of 5% oxygen concentration was reported to enhance cell growth, boost cell differentiation, and maintain its multipotency (Yang et al., 2012). The maintenance of stemness could be attributed to the activation of Notch downstream genes such as the *Hes* and *Hey* following upregulation of hypoxia-inducible factor (HIF-1) in a hypoxic condition (Sainson and Harris, 2006). Exposure to low oxygen concentration also reduces rate of telomere shortening and decreased incidence of double-strand breaks, leading to improved growth rate and genetic stability in adipose-derived MSCs (Estrada et al., 2012).

For induced pluripotent stem cell (iPSC) culturing, based on Yoshida et al., hypoxic environment (below 5% oxygen) can improve the efficiency of cell generation compared with cultivation in normal environment. When in hypoxia, it promotes the survival of neural crest cells and hematopoietic stem cells and prevents differentiation of human embryonic stem cells (Yoshida et al., 2009). Differentiation of human iPSCs to retinal progenitor cells was improved when the embryoid body was supplied with 2% oxygen. This condition activates the hypoxia-inducible factor, which in turn enhances both VEGF and LDHA gene expression. The authors also observed elevated expression levels of Six33 and Lhx2 implying early development of early eye field (Bae et al., 2012).

The above evidence demonstrates the significance of re-evaluating and opting for an optimal oxygen tension for incubation to grow the different types of normal stem cells without losing their intrinsic properties. Creating an environment that is similar to its origin where they usually reside (1%–9% O_2) rather than the common incubator condition set by the laboratory can improve the outcomes in clinical applications. For example, Kojima et al. reported that prior exposure of MSCs to hypoxic environment could enhance mouse liver regeneration following partial hepatectomy due to liver cirrhosis (Kojima et al., 2019). Given low oxygen tension, hematopoietic stem cells could grow rapidly and better engraft into bone marrow of a patient (Roy et al., 2012). This will certainly add value to the efficiency of treatment and reduce costs involved in delivering sufficient cell numbers for transplantation. Table 1 showed the advantages of using a hypoxic incubator for the culture of normal stem cells.

Three-dimensional (3D) organoid culture systems are another research area that should be explored as it allows the seeding cells to self-organize and recapitulate the developing organs. Hypoxic organoid culture is proven to be useful in modeling ischemic tissue injury, by exposing them to low oxygen concentration (1%). In a study done by Kim et al. (2021), brain organoids were created by using human skin fibroblast-induced neural stem cells and then incubated in hypoxic condition to establish an in vitro model of hypoxic brain injury. This model of tissue injury could be potentially used to elucidate molecular events in the pathogenesis, tissue repair, or developmental study, and for drug screening.

Besides looking into how the hypoxic microenvironment is able to play a role on normal stem cells, it is important to look at how they can influence the cancer stem cells too. In the study done by Heddleston et al. (2009), it showed that the hypoxic microenvironment (2% oxygen concentration) was able to maintain glioblastoma

Table 1 Review of the advantage using hypoxic incubator for stem cells and the type of stem cells can benefit from hypoxic culture.

Type of stem cells	Source	Suggestion level of oxygen tension	Benefits	Author year
Dental pulp stem cells	Dental pulp	5%	• Better morphology and smaller in size • Higher expression of stem cell markers • Higher migration capability toward G-CSF • High proliferation rate • Higher expression of angiogenic trophic factor (VEGF), neurotrophic factors (BDNF and NGF) • Decrease gene expression of Major Histocompatibility Complex Class-II • Enhance the trophic effect of the secretome	Ahmed et al. (2016)
Mesenchymal stem cell	N/A	2%–9%	• Maintain stem cell multipotency • Maintain genetic stability • Regulate the signaling of multiple cascades that promotes cell proliferation and chemotaxis • Reduce generation of reactive oxidative species (ROS) in the cells	Haque et al. (2013)
Mesenchymal stem cells	Adipose tissue	3%	• Increase growth rate • Increase lifespan by reducing the rate of telomere shortening per cell division by 24%, giving a better maintenance of telomere length. • Decrease reactive oxidative species (ROS) levels, thereby decreasing the incidence of double-strand breaks and limiting telomere erosion and aneuploidy. • Improve genetic stability	Estrada et al. (2012)
Mesenchymal stem cells	Umbilical cord	1%–10%	• Increase proliferation capacity • Reduce cell damage • Increase in glucose consumption and lactate production • Modulate the autocrine or paracrine activity of a variety of cytokines and growth factors	Lavrentieva et al. (2010)

Mesenchymal stem cells	Umbilical cord Wharton's Jelly	5%	• Enhance the human Wharton's Jelly-derived MSC secretome profiles when compared to the normoxic precondition • No changes in the neuroregulatory (neural differentiation) profile of the secretome	Teixeira et al. (2015)
Mesenchymal stem cells	Bone marrow	1%–7%	• Short-term proliferation • Long-term expansion efficiency • Maintain cell stemness • Modulate paracrine secretion capacity • Improve the expression of chemokine receptors, migration and engraftment ability	Hung (2013) and Tsai et al. (2012)
Induced pluripotent stem cell	Human dermal fibroblasts	5%	• Improve the efficiency of iPSCs generation • Induce cytotoxicity in human dermal fibroblasts	Yoshida et al. (2009)
Mesenchymal stem cells	Bone marrow	5%	• Promoted mouse liver regeneration after a partial hepatectomy	Kojima et al. (2019)
Adipose-derived stem cells	Adipose tissue	5%	• Enhance cell growth with a well-preserved differentiation potential • Boost cell proliferation • Effect of senescence may come into effect earlier than under standard growth conditions • Canonical ADcSC immunophenotype, expression of surface determinant remained stable • Maintain cell multipotency	Yang et al. (2012)
Hematopoietic stem cells	Umbilical cord blood	5%	• Increase cell number • Improve cell engraftment • Help to improve the recovery of colony-forming cells from normoxic culture	Roy et al. (2012)
Human induced pluripotent stem cells (iPSCs) MSUH001 and embryonic stem cells (hESCs) Shef3	Purchased from Spanish Stem Cell Bank, Madrid, Spain	2%	• Enhance generation of a specific population of progenitor cells in the central nervous system (CNS) • Increase differentiation efficiency of pluripotent stem cells to Retinal Progenitor Cells • Promote development into photoreceptors	Bae et al. (2012)

stem cells and to promote reprogramming of cancer cells toward a cancer stem cell phenotype. The same results were also observed by Murakami et al. (2014) in lung cancer stem cells. These studies suggest that hypoxia promotes tumor progression, metastasis, invasion, and therapeutic resistance in cancer biology. Therefore, fundamental cancer stem cell or anticancer drug targeting researches may employ the use of a hypoxic incubator to produce more accurate outcomes that could demonstrate higher relevance to human cancer biology and find an efficient treatment.

3. Procedural guide on choosing a hypoxic incubator

The purpose of the incubator is to create a suitable environment for your cell culture work. Most of the cell culture incubators will only monitor temperature, carbon dioxide, and humidity. The specific requirement for the cell culture work is mainly depending on the type and source of stem cell, and the model of treating disease.

Before you purchase the hypoxic incubator unit, it is important to find out the basic feature of an incubator for the type of your cell culture (Fig. 1) and if they will be used for basic research or manufacturing. Table 2 provides you some important questions for consideration before making a decision to purchase a cell culture incubator.

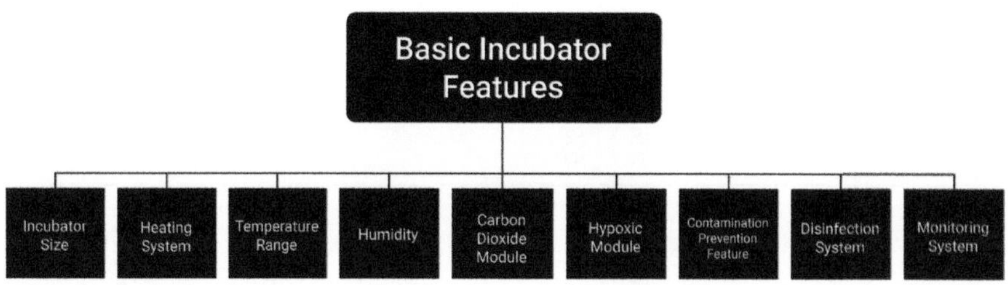

FIG. 1

Basic features of a cell culture incubator.

Table 2 Guide to assist you with the selection of the cell culture incubator.

No.	Incubator feature	Questions to ask yourself
1	Incubator constructive material	**(1)** What type of material is suitable to be placed in my research laboratory or manufacturing facility? – Example: Epoxy coating, powder coating, stainless steel
2	Incubator's chamber size	**(1)** What type of container do I use for my cell culture work? – Example: Cell culture flasks, multilayer culture flask, roller bottle, cell culture bag **(2)** What is the incubator chamber size required for my cell culture work?

Table 2 Guide to assist you with the selection of the cell culture incubator—cont'd

No.	Incubator feature	Questions to ask yourself
3	Incubator size	**(1)** What is the available space in the laboratory allocated for the incubator? **(2)** If I am constrained by the laboratory space, do I need a stacking system to place one incubator on above another?
4	Incubator support stand	**(1)** Do I place the incubator on the floor, table top or support roller stand?
5	Inner chamber heating system	**(1)** What is the incubator heating system used to distribute and maintain the inner chamber temperature? – Example: Water-jacket, direct heat, air jacket system
6	Inner door	**(1)** Is the temperature, carbon dioxide, or hypoxic level very critical to the cell culture? **(2)** If my incubator is placed in a general laboratory, do I need to reduce the rate of contamination in my incubator? **(3)** How many inner doors are needed for the chamber to maintain the inner chamber environment stability for the cell culture? – Example: 1, 2, 4, 6, or 8 doors
7	Environmental control system	**(1)** What type of control system should be used to control the incubator temperature, carbon dioxide, oxygen, and humidity level?
8	Sensor	**(1)** What type of carbon dioxide or oxygen sensor should be used? – Example: Infrared (IR), thermal conductivity (TC)
9	Hypoxic module	**(1)** Do I need a hypoxic module in addition to CO_2, temperature, and humidity for my cell culture work? **(2)** Do I need dual functions where I can interchange to either have hypoxic or nonhypoxic environment? **(3)** How many types of gases are required by the incubator to control the hypoxic environment? Note:It is important to check with your supplier to ensure that your laboratory are equipped with the gases that are required such as nitrogen, carbon dioxide, and with or without oxygen gas.
10	Decontamination system	**(1)** Does the incubator come with the contamination prevention feature for the chamber? – Example: High efficiency particulate air (HEPA) filter, ultra-violet (UV) light
11	Chamber cleaning procedure	**(1)** Is the chamber surface easy to clean with rounder corners, smooth surface and does not contain hidden corners?

Continued

Table 2 Guide to assist you with the selection of the cell culture incubator—cont'd

No.	Incubator feature	Questions to ask yourself
12	Self-disinfection system	**(1)** Do I need the incubator to be equipped with self-disinfection system? – Example: Moist heat, dry heat, hydrogen peroxide vapor
13	Continuous monitoring system	**(1)** Does my cell culture process need a continuous monitoring on the critical parameters such as temperature, carbon dioxide, oxygen and humidity? – Example: Online monitoring or in-build memory in the incubator
14	Independent monitoring probe	**(1)** If I need independent monitoring probe, is there a port available to insert the independent probe sensor?
15	Alarm system	**(1)** Do I need an alarm system to alert me immediately if the critical monitoring parameter value has dropped out from their allowable limit? **(2)** Which alarm method will be more effective method and can be integrated into my laboratory management? – Example: On site visual and sound alarm, SMS or email notification
16	Security system	**(1)** Do I need to set restricted access control for the incubator? – Example: Manual key system, password, fingerprint scanner, keycard reader.
17	Equipment qualification documentation	**(1)** Is the equipment qualification documentation needed by my laboratory facility or by the regulatory body? – Example: Installation qualification (IQ), Operation qualification (OQ)

4. Standard operation procedure/maintenance of a hypoxic incubator

In your laboratory, there should be a schedule and record for cleaning, service, and maintenance of the incubator. This is important to ensure the incubators are free from contamination and the internal compartments provide an optimal environment for cell cultivation. A daily, weekly, monthly, quarterly, or annual routine should be scheduled depending on your frequency of usage, external environment condition where you place your incubator, and policies or manufacture recommendation. One of the proposals for cleaning and maintenance provided by Eppendorf (Hamburg, Germany) (Hartmann and Jarvis, 2015) is as follows (Table 3):

Table 3 Proposals for cleaning and maintenance for the incubator.

Frequency	Details
Daily	**(1)** Inspect cell culture in the incubator daily. **(2)** If there is any spillage, remove and disinfect immediately with 70% ethanol or isopropanol.
Weekly	**(1)** Clean the water tray by discarding the existing water, disinfect the tray with 70% ethanol or isopropanol, and replace them with sterile water.
Monthly	**(1)** Empty the whole incubator chamber completely including the samples, racks, shelves, and doors. **(2)** Using lint free cloth, clean the interior chamber with soapy water, rinse with water followed by wiping the surface with 70% alcohol or stainless steel cleaning agent. **(3)** Disinfect the incubator if they are equipped with self-disinfection system. Please check with your manufacturer maintenance manual if there is a need to remove any of the incubator part (e.g., sensors and HEPA filter) before starting the disinfection program. Depending on the program, the disinfection process usually can take few hours to overnight. *Note:* **(1)** Steps 1 to 3 should be done: – for the new incubator unit, or if the incubator has not been operating for quite some time. – each time after completion of your culture work and before placing a new cell sample – if contamination of your cell culture is detected.
Every 6 months	**(1)** Replace the HEPA filter in your unit if the incubator is equipped with one.
Annually	**(1)** Service and maintenance of the incubator should be done at least once a year by an authorized service engineer. **(2)** The chamber environment parameters (e.g., temperature, O_2, and CO_2 gases) should also be validated once a year.

5. Conclusion

Based on all the articles reviewed, it is evident that hypoxic environment can bring benefits to the normal and cancer stem cell study. In this chapter, we have provided basic guidelines for further reference and consideration before making a decision to procure a cell culture incubator for research work or manufacturing processes. The optimal functions of an incubator can be prolonged if proper cleaning and maintenance are carried out throughout its lifespan and used with responsibility. Hopefully, this knowledge can provide benefits and solutions to improve some of the stem cell culturing techniques, which in turn may improve the treatment efficacy for tissue engineering and regenerative medicine.

Abbreviations

ADcSC	adipose-derived stem cell
Akt	Ak strain transforming
BDNF	brain-derived neurotrophic factor
CNS	central nervous system
CO2	carbon dioxide
CX3CR1	CX3C motif chemokine receptor 1
CXCR4	C-X-C chemokine receptor type 4
CXCR7	CXC chemokine receptor-7
G-CSF	granulocyte colony-stimulating factor
HEPA	high efficiency particulate air
Hes	hairy/enhancer of split
Hey	hairy/enhancer of split related with YRPW motif
HIF-1α	hypoxia-induced Factor
HO-1	heme oxygenase 1
Hsp70	70 kDa heat shock proteins
iPSC	induced pluripotent stem cell
IQ	installation qualification
IR	infrared
LDHA	lactate dehydrogenase A
Lhx2	lysyl hydroxylase 2
MSC	mesenchymal stem cells
NGF	nerve growth factor
02	oxygen
OQ	operation qualification
ROS	reactive oxidative stress
Six3	SIX homeobox 3
TC	thermal conductivity
UV	ultraviolet light
VEGF	vascular endothelial growth factor

References

Ahmed, N.E.M.B., Murakami, M., Kaneko, S., Nakashima, M., 2016. The effects of hypoxia on the stemness properties of human dental pulp stem cells (DPSCs). Sci. Rep. 6 (October), 1–10.

Bae, D., Mondragon-Teran, P., Hernandez, D., Ruban, L., Mason, C., Bhattacharya, S.S., Veraitch, F.S., 2012. Hypoxia enhances the generation of retinal progenitor cells from human induced pluripotent and embryonic stem cells. Stem Cells Dev. 21 (8), 1344–1355.

Estrada, J.C., Albo, C., Benguría, A., Dopazo, A., et al., 2012. Culture of human mesenchymal stem cells at low oxygen tension improves growth and genetic stability by activating glycolysis. Cell Death Differ. 19 (5), 743–755.

Haque, N., Rahman, M.T., Abu Kasim, N.H., Alabsi, A.M., 2013. Hypoxic culture conditions as a solution for mesenchymal stem cell based regenerative therapy. Sci. World J. 1–12.

Hartmann, I.K., Jarvis, J., 2015. CO2 incubators – best practices for selection, set-up and care. White Paper (Eppendorf) 29, 1–10.
Heddleston, J.M., Li, Z., McLendon, R.E., Hjelmeland, A.B., Rich, J.N., 2009. The hypoxic microenvironment maintains glioblastoma stem cells and promotes reprogramming towards a cancer stem cell phenotype. Cell Cycle 8 (20), 3274–3284.
Hung, S.C., 2013. Effects of hypoxic culture on bone marrow mesenchymal stem cells: from bench to bedside. Formos. J. Surg. 46 (2), 35–38.
Hung, S.C., 2016. Effects of hypoxic culture on bone marrow multipotent mesenchymal stromal cells. In: The Biology and Therapeutic Application of Mesenchymal Cells. John Wiley & Sons, pp. 196–207. 2(3).
Kim, M.S., Kim, D.H., Kang, H.K., Kook, M.G., Choi, S.W., Kang, K.S., 2021. Modeling of hypoxic brain injury through 3D human neural organoids. Cell 10 (2), 234.
Kojima, Y., Tsuchiya, A., Ogawa, M., et al., 2019. Mesenchymal stem cells cultured under hypoxic conditions had a greater therapeutic effect on mice with liver cirrhosis compared to those cultured under normal oxygen conditions. Regen. Ther. 11, 269–281.
Lavrentieva, A., Majore, I., Kasper, C., Hass, R., 2010. Effects of hypoxic culture conditions on umbilical cord-derived human mesenchymal stem cells. Cell Commun. Signal. 8, 1–9.
Murakami, A., Takahashi, F., Nurwidya, F., et al., 2014. Hypoxia increases Gefitinib-resistant lung cancer stem cells through the activation of insulin-like growth factor 1 receptor. PLoS One 9 (1), e86459.
Ortiz-Prado, E., Dunn, J.F., Vasconez, J., Castillo, D., Viscor, G., 2019. Partial pressure of oxygen in the human body: a general review. Am. J. Blood Res. 9 (1), 1–14.
Roy, S., Tripathy, M., Mathur, N., Jain, A., Mukhopadhyay, A., 2012. Hypoxia improves expansion potential of human cord blood-derived hematopoietic stem cells and marrow repopulation efficiency. Eur. J. Haematol. 88 (5), 396–405.
Sainson, R.C., Harris, A.L., 2006. Hypoxia-regulated differentiation: let's step it up a notch. Trends Mol. Med. 12 (4), 141–143.
Teixeira, F.G., Panchalingam, K.M., Anjo, S.I., et al., 2015. Do hypoxia/normoxia culturing conditions change the neuroregulatory profile of Wharton jelly mesenchymal stem cell secretome? Stem Cell Res. Ther. 6 (1), 1–14.
Tsai, C.C., Yew, T.L., Yang, D.C., Huang, W.H., Hung, S.C., 2012. Benefits of hypoxic culture on bone marrow multipotent stromal cells. Am. J. Blood Res. 2 (3), 148–159.
Wagner, W., Wein, F., Seckinger, A., Frankhauser, M., Wirkner, U., et al., 2005. Comparative characteristics of mesenchymal stem cells from human bone marrow, adipose tissue, and umbilical cord blood. Exp. Hematol. 33, 1402–1416.
Yang, S., Pilgaard, L., Chase, L.G., et al., 2012. Defined xenogeneic-free and hypoxic environment provides superior conditions for long-term expansion of human adipose-derived stem cells. Tissue Eng. Part C Methods 18 (8), 593–602.
Yoshida, Y., Takahashi, K., Okita, K., Ichisaka, T., Yamanaka, S., 2009. Hypoxia enhances the generation of induced pluripotent stem cells. Cell Stem Cell 5 (3), 237–241.

CHAPTER

Mouse embryonic stem cell-derived cerebral organoids

10

Nurfarhana Ferdaos[a] and John O. Mason[b]

[a]*Department of Pharmacology and Chemistry, Faculty of Pharmacy, Universiti Teknologi MARA Puncak Alam Campus, Selangor, Malaysia,* [b]*Simons Initiative for the Developing Brain and Centre for Discovery Brain Sciences, University of Edinburgh, Edinburgh, United Kingdom*

1. Introduction

A key property of embryonic stem (ES) cells that has made them attractive tools for many scientists is their pluripotency—the ability to give rise to a wide variety of specialized cell types, including neurons. Many different types of neurons have been successfully derived from mouse ES cells, including forebrain (Gaspard et al., 2009), midbrain (Jaeger et al., 2011) and cerebellum (Muguruma et al., 2010). ES cells are therefore useful tools for researchers interested in understanding how the brain develops normally during embryogenesis. It is important to understand this process because it will allow us to better understand the root causes of neurodevelopmental disorders which can arise when embryonic brain development goes wrong, ultimately leading to better ways to treat these disorders.

Until recently, most protocols for differentiating neurons from ES cells were based on 2D culture, in which neurons are grown adhering to a flat substrate, such as a tissue culture dish. These 2D protocols are efficient and useful, but, of course, normal brains develop in three dimensions, not two, so there are limits to the extent to which normal brain development can be studied in this way. This situation changed dramatically with the development of cerebral organoids, small tissue rudiments grown in vitro from stem cells that closely resemble embryonic brain tissue (Eiraku et al., 2008; Kadoshima et al., 2013; Lancaster et al., 2013; Mariani et al., 2012; Paşca et al., 2015) Cerebral organoids are surprisingly straightforward to grow—see the detailed protocol below. They contain many of the cell types found in the normal developing brain, arranged in the same way. They allow us, for the first time, to study how early brain structures develop in vitro.

The use of cerebral organoids has been extensively reviewed in the literature (for example: Benito Kwiecinski and Lancaster, 2019; Chan et al., 2020; Paşca, 2018). Cerebral organoids are set to transform the study of neural development. In this

Stem Cell Laboratory Techniques. https://doi.org/10.1016/B978-0-12-823729-8.00007-5

chapter, we highlight some of the key features that make cerebral organoids such useful tools in this field, concentrating on the mouse.

The embryonic brain in mice (as for all mammals) is highly inaccessible making it difficult to study directly. In contrast, as cerebral organoids are grown in vitro, they are highly accessible and cells within them can readily be visualized and their behaviors, such as interkinetic nuclear migration or cell migration tracked (Bershteyn et al., 2017; Karzbrun et al., 2018). Mouse brain development has been well studied, and we understand quite a lot about it. For an overview of early development of the mouse and human forebrain, including the differing cell types present and what we currently know about their normal roles, see (Mason and Price, 2016). In the normal developing cortex, the ventricular zone (VZ), located adjacent to the ventricles, contains progenitor cells known as radial glial cells (RGC) which proliferate rapidly, causing the brain to grow. RGC in mice can be identified by expression of the transcription factor Pax6. A second class of progenitor cells, known as basal progenitor cells, lack expression of Pax6 but do express Tbr2. These cells are located just outside the VZ. Each set of progenitors gives rise to neuronal daughter cells, which then migrate away from the VZ toward the outer, pial edge of the developing forebrain. Fig. 1 shows a set of immunostainings illustrating the locations of cells that express Emx1, Pax6, Tbr2 and Tbr1 in embryonic mouse brain at E12.5 and in cerebral organoids at a similar stage of development after 8 days in culture, demonstrating the close similarity.

Embryonic development of the brain is largely under genetic control, and ES cells are ideal for studying gene function as they are very straightforward to engineer, for example using CRISPR/Cas9 methods to introduce specific mutations to genes of interest. Furthermore, a vast array of genetically modified mouse strains have been generated over the last 30 years or as tools to investigate the mechanisms that control brain development and ES cells can readily be derived from any such line (Czechanski et al., 2014) and used to grow mutant organoids. One way that this is important is that it allows us to compare the phenotypes of specific mutations in mouse organoids with those found in mouse embryos carrying the same mutation, thus giving us confidence that the mechanisms by which organoids develop are the same as those used in vivo and further validating organoids as an experimental model system.

Many research groups have focussed on using human stem cells to grow cerebral organoids—this makes sense for those primarily interested in human brain development or who are studying disorders that affect human-specific developmental processes. Fewer studies employing organoids derived from mouse ES cells have been described so far; however, there are some situations in which mouse organoids have clear advantages as an experimental system (Marshall and Mason, 2019). Firstly, mouse and human organoids retain their normal in vivo growth rates, such that organoids grown from human stem cells take months to become mature, compared with 2 weeks or less for mouse cells. This speed makes mouse stem cell-derived organoids particularly suitable for protocol development. Optimizing organoid differentiation protocols can be tricky,

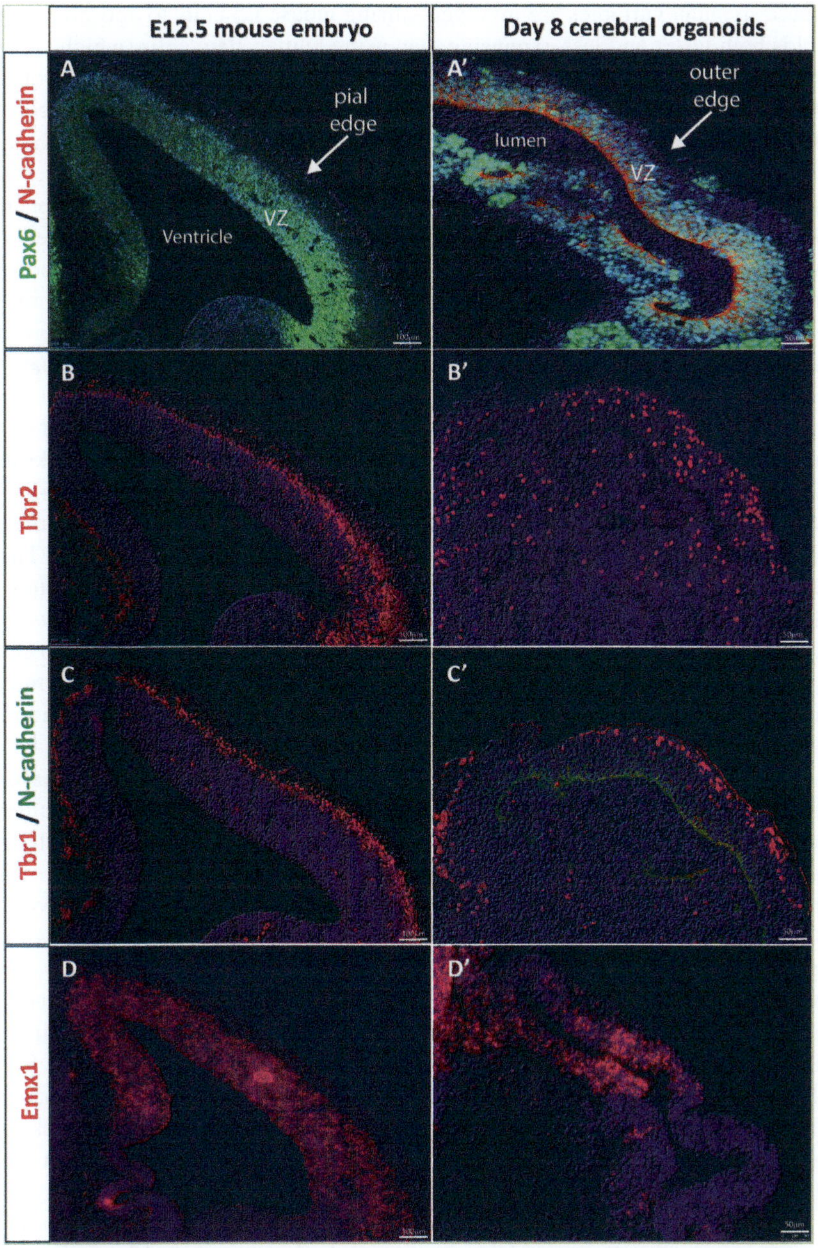

FIG. 1

Characteristics of cerebral organoids grown from mouse ES cells. Neuroepithelial-like structures of mouse cerebral organoids recapitulate embryonic neocortex as the cells express key cortical molecular markers in the correct pattern. They express Pax6 (A), an apical progenitor cell marker, Tbr2 (B), a basal progenitor cell marker, Tbr1 (C) a marker for early-born neurons (D), as well as Emx1 which is specifically expressed in embryonic cerebral cortex. *VZ*, ventricular zone.

because so many variables are involved—using mouse ES cells to optimize a protocol is both faster and less expensive than using human stem cells (which require more expensive reagents than mouse ES cells do). Once optimized, the protocol can then be adapted for use with human stem cells. For example, a protocol for growing thalamus organoids was originally devised using mouse ES cells (Shiraishi et al., 2017) then subsequently adapted to human (Xiang et al., 2019). Secondly, as mentioned above, there is a wealth of in vivo mutant phenotype data from mice for comparison purposes, giving confidence that the effects of mutations on organoids are the same as those seen in animals. A further important point to note about mouse organoids is that they should lead to a reduction in the numbers of mice used by scientists for their experiments—organoids have the potential to replace mice as an experimental model in many situations (Marshall and Mason, 2019).

While it is already clear that organoids have great potential as models to study brain development, it is important to remember that they have some drawbacks too. One notable limitation is that organoids often display high levels of variability—differentiation is often not uniform, such that organoids will contain regions of well-organized neural tissue but also some regions that have different fates. There is also often variability between organoids, such that the extent and efficiency of neural differentiation can vary between individual organoids in a batch, as well as between batches. This is important to bear in mind when using organoids to characterize the phenotype of a mutation—it is essential to use a quantitation strategy that can clearly distinguish the phenotype from inherent variation between organoids. Organoids are an incomplete representation of normal brain tissue—while neural cells closely resemble those found in vivo, organoids lack other specialized cell types such as microglia and blood vessels. The absence of a blood supply means that cells in the center of the organoid may not obtain sufficient nutrients from the medium, likely inhibiting growth and sometimes leading to their death and the formation of a core of necrotic cells. This limitation may be circumvented by slicing the organoids open to allow culture medium to reach the center of the organoids (Qian et al., 2020). Finally, while mouse organoids are likely good models of mouse brain development, they lack human-specific features, such as the outer subventricular zone, and therefore have limitations in terms of their suitability to study at least some aspects of normal human brain development. Similarly, mouse organoids are less suitable than human organoids for modeling some neurodevelopmental disorders, because of interspecies differences; for example in progenitor subtypes which are present in human but not in mice (Mason and Price, 2016).

Because fewer studies involving mouse organoids have been published, they are in general less well characterized than human cerebral organoids and fewer protocols have been described. In this chapter, we provide a set of detailed protocols for the production of cortical cerebral organoids from mouse ES cells. The method is closely based on the SFEBq method developed in the Sasai lab (Eiraku et al., 2008; Nasu et al., 2012; Watanabe et al., 2005). The method is summarized in Fig. 2. In outline,

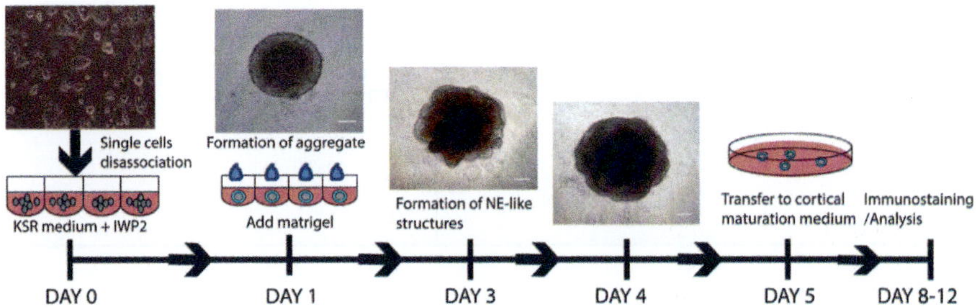

FIG. 2

Overview of the protocol. ~5000 dissociated ES cells are aggregated quickly in 96 well plates in medium containing the Wnt inhibitor IWP2. One day later, the extracellular matrix preparation Matrigel is added to promote epithelial formation. At 4 days, they are transferred to cortical maturation medium then left to differentiate for the required time.

the method relies on the tendency of differentiating ES cells to adopt anterior forebrain fate by default, so long as they are protected from the influence of posteriorizing factors, such as Wnts (Wilson and Houart, 2004). There are two undefined components in the differentiation medium, KSR (knockout serum replacement) and Matrigel, providing possible sources of variation. Defined, synthetic substitutes for Matrigel are being developed (Ranga et al., 2016), possibly leading to reduced variation in future. We do not routinely agitate our cultures, but this has been found to be important for growing human organoids (Lancaster et al., 2013) and may lead to improved nutrient exchange and therefore organoid growth.

2. Methods

2.1 Preparation of media and reagents

2.1.1 Media

Embryonic stem cell medium

	500 mL
Glasgow minimum essential medium	1 ×
Sodium pyruvate	1 mM
Glutamax	2 mM
Nonessential amino acids	1 ×
Beta-mercaptoethanol	0.1 mM
Fetal calf serum	10%
Leukemia inhibitory factor	1000 U/mL

Freezing medium

	10 mL
Embryonic stem cell medium	1 ×
FCS	10%
DMSO	10%

Knockout serum free medium

	50 mL
GMEM	1 ×
Pyruvate	1 mM
Nonessential amino acid	1 ×
Beta-mercaptoethanol	0.1 mM
Knockout serum replacement	10%

Cortical maturation medium

	50 mL
DMEM/F12, 50 mL	1 ×
Glutamax	1 mM
N2, 100 ×	1 ×

2.1.2 Other reagents

Enzymes

TryPLE (recombinant enzyme, a trypsin replacement)
Extracellular matrix
Matrigel (growth-factor reduced)

	1%
Porcine gelatin powder	1 g
PBS	1 ×

Mix with heat and filter by using 0.2 μm filter. Dilute 5 mL of 1% porcine gelatin in 45 mL PBS to make up 0.1% and store at 4°C.

	4%
Paraformaldehyde	20 g
PBS	500 mL

Mix and heat until PFA dissolves and solution becomes clear. Aliquot into 50 mL Falcon tubes and store at −20°C.

	30%
Sucrose	15 g
PBS	50 mL

Mix with moderate heating. Aliquot into 50 mL Falcon tubes and store at −20°C.

Embedding medium

Mix 30% sucrose and OCT in 50:50 ratio. To make 50 mL embedding medium, mix 25 mL of 30% sucrose and 25 mL OCT.

2.2 Mouse embryonic stem cell culture

2.2.1 Maintenance of embryonic stem cells

Mouse embryonic stem (ES) cells are propagated in ES medium on 25 cm^2 tissue culture flasks at 37°C, 5% CO_2. Healthy ES cell cultures require subculturing every 2 days, by splitting subconfluent cells at a ratio of 1:10.

2.2.2 Subculture

Upon 70%–80% confluence, passage cells by gentle trypsinization:

- **i.** Coat 25 cm^2 flask with 0.1% gelatin (1 mL) for 30 min. Aspirate gelatin prior to plating
- **ii.** Wash cells with PBS, twice
- **iii.** Add 1 mL TryPLE and allow the cells to detach by leaving them in the incubator (37°C, 5% CO_2), 2–3 min
- **iv.** Add 9 mL ES medium and further disassociate cells by gentle trituration
- **v.** Split cells in 1:10 (1 mL from the 10 mL cell suspension) into a new 25 cm^2 flask, add 4 mL fresh ES medium
- **vi.** Incubate cells at 37°C, 5% CO_2

2.2.3 Cryopreservation and thawing

To stock cells:

i. Aspirate medium and wash cells with PBS, twice
ii. Add 1 mL TryPLE and allow the cells to detach by leaving them in the incubator (37°C, 5% CO_2), 2–3 min
iii. Add 9 mL ES medium and further disassociate cells by gentle trituration
iv. Centrifuge at 250 × g, 5 min to pellet cells
v. Decant the medium and add 1 mL fresh freezing medium
vi. Transfer into 1.5 mL cryovial
vii. Keep in −80°C overnight to allow the cells to freeze slowly
viii. Store in liquid nitrogen tank for long-term storage

To revive cryopreserved ES cells:

i. Coat 25 cm^2 flask with 0.1% gelatin (1 mL) for 30 min. Aspirate gelatin prior to plating
ii. Retrieve frozen ES cells from liquid nitrogen tank and thaw quickly in 37°C waterbath
iii. Transfer the cells to centrifuge tube in 9 mL ES medium and spin at 250 × g, 5 min
iv. Decant the medium and resuspend the cells in 5 mL fresh ES medium
v. Plate the cells on gelatin-coated 25 cm^2 flask
vi. Change medium the day after

2.3 Cerebral organoids 3-dimensional (3D) culture

Stage 1: Day 0

i. Grow mouse ES cells until 70%–80% confluent
ii. Aspirate medium and wash cells with PBS, twice
iii. Add 1 mL TryPLE and incubate cells in 37°C, 5% CO_2, 2–3 min
iv. Add 5 mL ES medium and triturate cells gently to facilitate single cell suspension
v. Spin at 250 × g, 10 min to pellet cells
vi. Decant the supernatant and resuspend the cell pellet in KSR medium
vii. Count cells and determine cell concentration
viii. Dilute cells in KSR medium and add 2.5 μM IWP2
ix. Aliquot 0.1 mL (5000 cells) into 96-U shaped well plates
x. Incubate at 37°C, 5% CO_2

Stage 2: Day 1

i. The cells aggregate spontaneously overnight to form a single sphere
ii. Thaw Matrigel on ice and use precooled tips (−20°C) to prevent solidification
iii. Add 200 μg/mL Matrigel onto the aggregates
iv. Incubate at 37°C, 5% CO_2

Stage 3: Day 5

i. On day 5, aggregates form bright and radial neuroepithelial-like structures, identified as cerebral organoids. Using a Pasteur pipette, transfer up to 16 cerebral organoids to a 50 mm bacterial grade Petri dish in 10 mL cortical maturation medium
ii. Change medium every other day, by carefully removing half of the medium (5 mL) using a P1000 pipette and replenish with an equal volume of fresh medium
iii. Leave cerebral organoids at 37°C, 5% CO_2 until the required stage of development

2.4 Tissue processing

To characterize the structure of cerebral organoids, they are usually cryosectioned. Prior to sectioning, the tissues must be processed and preserved.

2.4.1 Fixation

i. Transfer organoids from the Petri dish into 12-well plates, using a 1 mL Pasteur pipette
ii. Aspirate residual medium and wash the organoids with PBS, twice
iii. Add 3 mL of 4% PFA and incubate at room temperature for 20 min, on a rocking platform
iv. Carefully remove the PFA using a P200 pipette and discard the PFA safely
v. Rinse organoids with PBS three times, 10 min each
vi. Keep in PBS and store at 4°C until further use

2.4.2 Cryoprotection, embedding and cryosectioning

i. Immerse fixed organoids in 30% sucrose at 4°C, overnight
ii. Aspirate sucrose solution using a P200 pipette
iii. Immerse organoids in embedding medium at 4°C, 1 h, on a rocking platform
iv. Transfer up to 12 organoids into a plastic mold. Allow them to settle to the bottom by leaving them undisturbed for 10 min
v. Aspirate residual medium using a P200 pipette and arrange the organoids to the center of the mold
vi. Immediately place the mold on a dry ice to fix the organoids in position at the center and quickly fill half of the mold with embedding medium
vii. Allow the blocks to freeze completely by leaving the mold in a box of dry ice, around 15 min
viii. Store at −80°C until further use
ix. Cut the blocks at 10 μm thickness in a cryostat chamber and transfer sections to superfrost slides
x. Store slides at −20°C or −80°C until further use

2.5 Sources of materials

Item	Catalog number	Supplier
Glasgow minimum essential medium	21710-025	Invitrogen (Thermo Fisher Scientific)
100 mM sodium pyruvate	11360-039	
Glutamax	35050061	
100 × nonessential amino acids	11140-035	
50 mM B-mercaptoethanol	31350-010	
DMEM/F12	21331-020	
100 × N2	17502-048	
Knockout serum replacement (KSR)	10828010	
Fetal calf serum (FCS)		
Matrigel (growth-factor reduced)	356231	Corning
IWP2	I0536-5MG	Sigma
PrimeSurface 96U plate	MS-9096U	Alpha Laboratories Ltd

References

Benito Kwiecinski, S., Lancaster, M., 2019. Brain organoids: human neurodevelopment in a dish. Cold Spring Harb. Perspect. Biol. 12, a035709. https://doi.org/10.1101/cshperspect.a035709.

Bershteyn, M., Nowakowski, T., Pollen, A., Di Lullo, E., Nene, A., Wynshaw-Boris, A., Kriegstein, A., 2017. Human iPSC-derived cerebral organoids model cellular features of lissencephaly and reveal prolonged mitosis of outer radial glia. Cell Stem Cell 20. https://doi.org/10.1016/j.stem.2016.12.007.

Chan, W.K., Griffiths, R., Price, D., Mason, J., 2020. Cerebral organoids as tools to identify the developmental roots of autism. Mol. Autism 11. https://doi.org/10.1186/s13229-020-00360-3.

Czechanski, A., Byers, C., Greenstein, I., Schrode, N., Donahue, L., Hadjantonakis, A.-K., Reinholdt, L., 2014. Derivation and characterization of mouse embryonic stem cells from permissive and nonpermissive strains. Nat. Protoc. 9, 559–574. https://doi.org/10.1038/nprot.2014.030.

Eiraku, M., Watanabe, K., Matsuo-Takasaki, M., Kawada, M., Yonemura, S., Matsumura, M., Wataya, T., Nishiyama, A., Muguruma, K., Sasai, Y., 2008. Self-organized formation of polarized cortical tissues from ESCs and its active manipulation by extrinsic signals. Cell Stem Cell 3, 519–532. https://doi.org/10.1016/j.stem.2008.09.002.

Gaspard, N., Bouschet, T., Herpoel, A., Naeije, G., van den Ameele, J., Vanderhaeghen, P., 2009. Generation of cortical neurons from mouse embryonic stem cells. Nat. Protoc. 4, 1454–1463. https://doi.org/10.1038/nprot.2009.157.

Jaeger, I., Arber, C., Risner-Janiczek, J., Kuechler, J., Pritzsche, D., Chen, I.-C., Naveenan, T., Ungless, M., Li, M., 2011. Temporally controlled modulation of FGF/ERK signaling directs midbrain dopaminergic neural progenitor fate in mouse and human pluripotent stem cells. Development 138, 4363–4374. https://doi.org/10.1242/dev.066746.

Kadoshima, T., Sakaguchi, H., Nakano, T., Soen, M., Ando, S., Eiraku, M., Sasai, Y., 2013. Self-organization of axial polarity, inside-out layer pattern, and species-specific progenitor dynamics in human ES cell-derived neocortex. Proc. Natl. Acad. Sci. U. S. A. 111. https://doi.org/10.1073/pnas.1315710110.

Karzbrun, E., Kshirsagar, A., Cohen, S., Hanna, J., Reiner, O., 2018. Human brain organoids on a chip reveal the physics of folding. Nat. Phys. 14. https://doi.org/10.1038/s41567-018-0046-7.

Lancaster, M., Renner, M., Martin, C.-A., Wenzel, D., Bicknell, L., Hurles, M., Homfray, T., Penninger, J., Jackson, A., Knoblich, J., 2013. Cerebral organoids model human brain development and microcephaly. Nature 501. https://doi.org/10.1038/nature12517.

Mariani, J., Simonini, M., Palejev, D., Tomasini, L., Coppola, G., Szekely, A., Horvath, T., Vaccarino, F., 2012. Modeling human cortical development in vitro using induced pluripotent stem cells. Proc. Natl. Acad. Sci. U. S. A. 109, 12770–12775. https://doi.org/10.1073/pnas.1202944109.

Marshall, J., Mason, J., 2019. Mouse vs man: organoid models of brain development & disease. Brain Res. 1724, 146427. https://doi.org/10.1016/j.brainres.2019.146427.

Mason, J., Price, D., 2016. Building brains in a dish: prospects for growing cerebral organoids from stem cells. Neuroscience 334. https://doi.org/10.1016/j.neuroscience.2016.07.048.

Muguruma, K., Nishiyama, A., Ono, Y., Miyawaki, H., Mizuhara, E., Hori, S., Kakizuka, A., Obata, K., Yanagawa, Y., Hirano, T., Sasai, Y., 2010. Ontogeny-recapitulating generation and tissue integration of ES cell-derived Purkinje cells. Nat. Neurosci. 13, 1171–1180. https://doi.org/10.1038/nn.2638.

Nasu, M., Takata, N., Danjo, T., Sakaguchi, H., Kadoshima, T., Futaki, S., Sekiguchi, K., Eiraku, M., Sasai, Y., 2012. Robust formation and maintenance of continuous stratified cortical neuroepithelium by laminin-containing matrix in mouse ES cell culture. PLoS One 7, e53024. https://doi.org/10.1371/journal.pone.0053024.

Paşca, S., 2018. The rise of three-dimensional human brain cultures. Nature 553, 437–445. https://doi.org/10.1038/nature25032.

Paşca, S., Sloan, S., Clarke, Tian, Y., Makinson, C., Huber, N., Kim, T., Park, J.-Y., O'Rourke, N., Nguyen, K., Smith, S., Huguenard, J., Geschwind, D., Barres, B., 2015. Generation of functional cortical neurons and astrocytes from human pluripotent stem cells in 3D cultures. Nat. Methods 12. https://doi.org/10.1038/nmeth.3415.

Qian, X., Su, Y., Adam, C., Deutschmann, A., Pather, S., Goldberg, E., Su, K., Li, S., Lu, L., Jacob, F., Nguyen, P., Huh, S., Hoke, A., Swinford-Jackson, S., Wen, Z., Gu, X., Pierce, R., Wu, H., Briand, L., Ming, G., 2020. Sliced human cortical organoids for Modeling distinct cortical layer formation. Cell Stem Cell 26. https://doi.org/10.1016/j.stem.2020.02.002.

Ranga, A., Girgin, M., Meinhardt, A., Eberle, D., Caiazzo, M., Tanaka, E., Lutolf, M., 2016. Neural tube morphogenesis in synthetic 3D microenvironments. Proc. Natl. Acad. Sci. U. S. A. 113. https://doi.org/10.1073/pnas.1603529113.

Shiraishi, A., Muguruma, K., Sasai, Y., 2017. Generation of thalamic neurons from mouse embryonic stem cells. Development 144, dev.144071. https://doi.org/10.1242/dev.144071.

Watanabe, K., Kamiya, D., Nishiyama, A., Katayama, T., Nozaki, S., Kawasaki, H., Watanabe, Y., Mizuseki, K., Sasai, Y., 2005. Directed differentiation of telencephalic precursors from embryonic stem cells. Nat. Neurosci. 8, 288–296. https://doi.org/10.1038/nn1402.

Wilson, S., Houart, C., 2004. Early steps in the development of the forebrain. Dev. Cell 6, 167–181. https://doi.org/10.1016/S1534-5807(04)00027-9.

Xiang, Y., Tanaka, Y., Çakir, B., Patterson, B., Kim, K.-Y., Sun, P., Kang, J., Zhong, M., Liu, X., Patra, P., Lee, S.-H., Weissman, S., Park, I.-H., 2019. hESC-derived thalamic organoids form reciprocal projections when fused with cortical organoids. Cell Stem Cell. https://doi.org/10.1016/j.stem.2018.12.015.

CHAPTER

Standardized practice of stem cell production

11

Baskar Subramani[a], Aida Suraya Binti Abu Yazid[b], and Mathanakumara Ealam Selvan[b]

[a]*Medixcell Sdn Bhd, Bandar Sg Long, Kajang, Malaysia,* [b]*Faculty of Science, University of Malaya, Kuala Lumpur, Malaysia*

1. Cell line identification and authentication by short tandem repeat (STR) DNA profiling

1.1 Background and principle

Short tandem repeat (STR) analysis is considered as the gold standard for the authentication of human cell lines. The method is straightforward, economical, and dependable for the detection and intraspecific identification of cross-contaminating cells. A good management plan for cells during expansion and analyses must be executed, such as the establishment of Working Cell Bank and Master Cell Bank. Cell suspension at the volume of 20.0 μL (2.0×10^5 cells) should be spotted onto Flinders Technology Associates (FTA) paper immediately after delivery or thawing (if the cells were cryopreserved) in order to create the baseline STR profile for analysis. As a result, all subsequent STR DNA profiles carried out on the various cell banks will be compared to the initial baseline STR profile from the first donor material. The Master Cell Bank and Working Cell Bank can be created by multiplying the donor's remaining cells. For quality assurance and quality control, a representative sample from each bank must undergo a second STR DNA profile analysis after expansion. Then, if there is a match of more than 80% between the STR DNA profiles from the Master and Working banks and the baseline profile of the original donor material, the cells are deemed to be authentic.

In order to determine gender, STR DNA profile analysis simultaneously amplifies 17 STR markers and amelogenin. This technique has a 1.0×10^{-18} level of discrimination between human cell lines. The PCR amplicons are then separated using capillary electrophoresis utilizing internal size standards (ISS). The attributed alleles are converted to numerical values to create a baseline profile after being compared to the allelic ladders for the sized pieces. The reference database is created utilizing the DNA baseline profiles. The reference database's baseline profile of the particular cell line is then utilized as a standard for all STR DNA profile tests that will

Stem Cell Laboratory Techniques. https://doi.org/10.1016/B978-0-12-823729-8.00011-7

subsequently be carried out on the numerous cell banks. The reference database's other cell lines must be compared to the STR DNA profile in order to be meaningful.

1.2 Protocol

Materials and apparatus

1. DMEM +10% FBS (D10) (Gibco; Thermo Fisher Scientific Inc., Massachusetts, United States)
2. 2800 M Control DNA (Promega Corporation; Wisconsin, United States)
3. Phosphate-buffered saline (PBS) (Gibco; Thermo Fisher Scientific Inc., Massachusetts, United States)
4. Dry heating block
5. 0.05% Trypsin-EDTA (Gibco; Thermo Fisher Scientific Inc., Massachusetts, United States)
6. Water bath
7. 0.4% Trypan blue (Gibco; Thermo Fisher Scientific Inc., Massachusetts, United States)
8. Ice-water bath or crushed ice
9. Cell counter/hemocytometer
10. Centrifuge compatible with 96-well plates
11. FTA Card (Qiagen; Hilden, Germany)
12. Aerosol-resistant pipet tips
13. Centrifuge
14. 36-cm 3500/3500xL capillary array (Applied Biosystems; Massachusetts, United States)
15. PowerPlex 18D System (Promega Corporation; Wisconsin, United States)
16. 96-Well retainer and base set (standard) (Applied Biosystems; Massachusetts, United States)
17. GeneAmp PCR System 9700 thermal cycler (Applied Biosystems; Massachusetts, United States) or equivalent
18. POP-4 polymer in a pouch for the Applied Biosystems 3500 or 3500xL genetic Analyzer (Applied Biosystems; Massachusetts, United States)
19. MicroAmp optical 96-well reaction plate (Applied Biosystems; Massachusetts, United States) or equivalent
20. Anode buffer container (Applied Biosystems; Massachusetts, United States)
21. Microcentrifuge
22. Cathode buffer container (Applied Biosystems; Massachusetts, United States)
23. Conditioning reagent pouch for the Applied Biosystems 3500 or 3500xL genetic Analyzer (Applied Biosystems; Massachusetts, United States)
24. 1.2 mm Harris Micro-Punch or equivalent
25. Hi-Di formamide (Applied Biosystems; Massachusetts, United States)
26. Cutting mat
27. PowerPlex 5-Dye Matrix Standards, 3100/3130 (Promega Corporation; Wisconsin, United States)
28. PowerPlex D 5× master Mix (Promega Corporation; Wisconsin, United States)
29. CC5 Internal Lane Standard 500 (Promega Corporation; Wisconsin, United States)
30. PowerPlex 18D 5× Primer Pair mix (Promega Corporation; Wisconsin, United States)
31. PowerPlex 18D Allelic Ladder Mix (Promega Corporation; Wisconsin, United States)
32. Applied Biosystems 3500 or 3500xL Genetic Analyzer (Applied Biosystems; Massachusetts, United States)

Note: The Applied Biosystems 3130, 3130xL, 3500, and 3500xL Genetic Analyzers and the ABI PRISM 3100 and 3100-Avant are both compatible with the PowerPlex 18D System from Promega Corporation. The manufacturer's instructions should be consulted for full instructions.

1.2.1 Cell harvest

Both forms of DNA may be used; thus, either the cell suspension can be immediately spotted onto FTA paper or it can be used to extract liquid DNA for STR DNA profile analysis. If the harvested cells are to be used for liquid DNA extraction (total DNA/genomic DNA), and with the assumption that a commercial DNA extraction kit is employed, the instructions provided with the kit should be followed according to the manufacturer. The following is the general protocol for cell harvest; the exact volume of reagents depends on the size of the culture flask/plate used.

1. Remove spent/used medium.
2. Use PBS to gently swirl on the surface of the cells after adding it. PBS should be drained and discarded afterwards.
3. Cover the cell surface with 0.05% trypsin-EDTA and incubate for 2–5 min, or until all of the cells have detached from the flask's surface.
4. Stop trypsinization by adding two times the volume of 0.05% trypsin-EDTA added in step 3 (i.e., add 6.0 mL of D10 if 3.0 mL trypsin/EDTA was added). Gently resuspend cells in the flask and transfer suspension into a conical tube (50-mL or 15-mL conical tube according to volume collected).
5. Centrifuge cells at for 10 min at 200 × g. Dispose the resulting supernatant.
6. After draining the supernatant, use PBS to thoroughly wash the cell pellet twice.
7. Rehydrate the cell pellet with a small amount of PBS. Calculate the proportion of viable cells after doing a cell count with 0.4% Trypan blue.
8. Spread the cell suspension out onto FTA paper or for DNA isolation.

1.2.2 Spotting cells onto FTA card

1. Spot about 20.0 μL of PBS containing approximately 2.0×10^5 suspension cells onto an FTA Card and allow to air-dry. *Note*: Cards can be kept in a cold, dry environment for future usage or long-term archiving.
2. In the center of the sample spot, punch a 1.2-mm disk.
3. Eject the disk into the proper reaction plate well using a plunger.

1.2.3 Amplification setup

1. Defrost the PowerPlex 18D 5× Primer Pair mix and PowerPlex D 5× master Mix.
2. Determine how many responses are required for the run based on the number of samples. Include pipetting error corrections and controls (both positive and negative) in the calculation.
3. Mark a clear MicroAmp plate with a 0.2 mL capacity for reaction assembly.
4. Table 1 below is a guide on the calculation of volume of reagents needed for preparation of PCR amplification mix. Fill in the number of samples and add reagents according to the column labeled "calculated volume" below to a sterile tube.
5. For 10 s, vortex the PCR mixture.

Table 1 Calculation guide for volume of reagents needed for preparation of PCR amplification Mix.

Components for PCR amplification mix	Volume per each reaction (μL)	×	Number of samples/ reactions	Calculated volume (μL)
PowerPlex D 5× Master Mix	5.0	×	___	___
PowerPlex 18D 5× Primer Pair mix	5.0	×	___	___
Water (PCR grade)	15.0	×	___	___
Total Volume	**25.0**	×	___	___

6. Insert 25.0 μL of the PCR amplification mix into every reaction tube containing an FTA disk.
7. For positive control of the amplification:
   a. Vortex the 2800 M Control DNA tube.
   b. Reduce the concentration of 2800 M Control DNA to 5.0 ng/μL.
   c. Diluted Control DNA should be pipetted at the volume of 1.0 μL into reaction well which hold 25.0 μL of PCR amplification mix prepared.
   d. Use a reaction well containing either 25.0 μL amplification mix without DNA or a reaction well containing 25.0 μL amplification mix with FTA disk without DNA to perform a negative control for the amplification.
8. To get the disks to settle at the base of the wells, seal the plate, and quickly centrifuge it (quick spin).

1.2.4 Thermal cycling

1. In the thermal cycler, place the MicroAmp plate.
2. Clarify and start the protocol as recommended in the Table 2 below.

1.2.5 Using the applied biosystems 3500 or 3500xL genetic analyzer for detection of amplified fragments

The PowerPlex 18D primers are covalently linked to fluorescent molecules. In a single PCR, different loci will be analyzed using several sets of primers with varying fluorescent "colors." After PCR, the reaction mixture is supplemented with internal size standards, and the DNAs are then separated by capillary gel electrophoresis

Table 2 Recommended protocol for PCR Run (thermocycler setup).

Steps	Temperature (°C)	Duration	Cycles
1	96.0	2 min	
2	94.0	10 s	27
3	60.0	1 min	
4	60.0	20 min	

according to size. After the data have been collected, the amplicons will be sorted using an internal size criterion by the application GeneMapper ID-X. STR genotyping of cell lines will be performed by translating the sizes of amplicons to alleles and contrasting them with allelic ladders. The supplies and reagents that can be used in this test are listed below.

1. Dry heating block **2.** Water bath **3.** Crushed ice/ice-water bath **4.** Centrifuge compatible with 96-well plates **5.** Aerosol-resistant pipette tips **6.** 36-cm 3500/3500xL capillary array (Applied Biosystems; Massachusetts, United States) **7.** 96-Well retainer and base set (standard) (Applied Biosystems; Massachusetts, United States) **8.** POP-4 polymer in a pouch for the Applied Biosystems 3500 or 3500xL genetic Analyzer (Applied Biosystems; Massachusetts, United States)	**9.** Anode buffer container (Applied Biosystems; Massachusetts, United States) **10.** Cathode buffer container (Applied Biosystems; Massachusetts, United States) **11.** Conditioning reagent pouch for the Applied Biosystems 3500 or 3500xL genetic Analyzer (Applied Biosystems; Massachusetts, United States) **12.** Hi-Di formamide (Applied Biosystems; Massachusetts, United States) **13.** PowerPlex 5-Dye Matrix Standards, 3100/3130 (Promega Corporation; Wisconsin, United States)

Preparation of sample

1. To prepare the loading cocktail, combine the following amounts of CC5 Internal Lane Standard 500 (CC5 ILS 500) and Hi-Di formamide:
   a. Amount of loading cocktail = (1.0 µL CC5 ILS 500 × number of injections) + (10.0 µL Hi-Di formamide × number of injections) * The internal lane standard loading quantity may be modified as necessary based on how intense the size standard band is.
2. Vortex to combine the loading cocktail.
3. Pipette the loading cocktail into every well at the volume of 11.0 µL.
4. Add either 1.0 µL PowerPlex 18D Allelic Ladder Mix or 1.0 µL of amplified sample accordingly to the assigned wells.
5. Use the proper septa to cover the wells.
6. Centrifuge the plate for a short spin to eliminate any pockets of air from the mixture in the wells.
7. Heat the samples at 95°C for 3 min for denaturation before to loading the device.
8. Cool the dish at once in an ice bath or with crushed ice for 3 min.

Instrument preparation and use

With the Applied Biosystems 3500 or 3500xL Genetic Analyzer, amplified fragments can be located. You may find instructions on how to operate and maintain the instrument in the User Guide (https://www.thermofisher.com/document-connect/document-connect.html?url=https://assets.thermofisher.com/TFS-Assets%2FLSG%2Fmanuals%2F100031809_3500_3500xL_Software_v3_1_UG.pdf) and

Promega (https://worldwide.promega.com/resources/protocols/#sort=relevancy) can be referred for a detailed method for data analysis.

1.2.6 Analysis of data

PowerPlex 18D Panel, Bins, Stutter Text Files, and GeneMapper ID-X Software, version 1.2, are the programs that may be utilized for data analysis. Visit www.promega.com/geneticidtools/panels_bins/ to get panel and bin text files for GeneMapper ID-X software's automated genotype assignment for analyzing data produced by the PowerPlex 18D System.

1.2.7 Interpretation of data

The analyzed data are then compared with original sample/reference sample. A similarity match of ≥80% between analyzed sample and its reference is considered as authentic. Overall, the interpretation of percentage of similarity is as below:

Authentication of cell line from STR profile:

- **Authentic**: ≥80% match between cell line and its original tissue or its derivatives
- **Unrelated**: ≤56% match between cell line and its original tissue or its derivatives
- **Re-test/further analysis**: match profile between 56% and 80%

For one human cell line to be distinguished from another, a minimum of 8 core STR loci are needed. vWA, Th01, TPOX, CSF1PO, D5S818, D13S317, D7S820, and D16S539 are the 8 loci in question. Fig. 1 demonstrates that whereas cell lines derived from various organs from the same individual have varied STR profiles, same STR profile is observed among unique cell lines (Fig. 2).

2. Human leukocyte antigen (HLA) analysis procedure—Polymerase chain reaction-based sequence-specific primers (PCR-SSP) method

2.1 Background and principle

By enabling the immune system to discriminate between "self" and "nonself," the HLA system; a complex, codominantly inherited antigen system, is crucial to maintaining human health. Regarding the compatibility of the donor and recipient, HLA is one of the most crucial criteria impacting the success of organ transplantation. Therefore, the determination of individual HLA antigen combinations is utilized as a basis for selection for donor and recipient.

There are many emerging applications of molecular techniques for HLA typing to improve transplantation success and recovery. Among commonly employed genetic methods of HLA typing include polymerase chain reaction-based sequence-specific primers (PCR-SSP), polymerase chain reaction-based sequence-specific oligonucleotide probe method (PCR-SSO), and sequencing-based typing (SBT).

The polymerase chain reaction-based sequence-specific primers (PCR-SSP) method is often applied because the technique is simple, rapid, reproducible as well

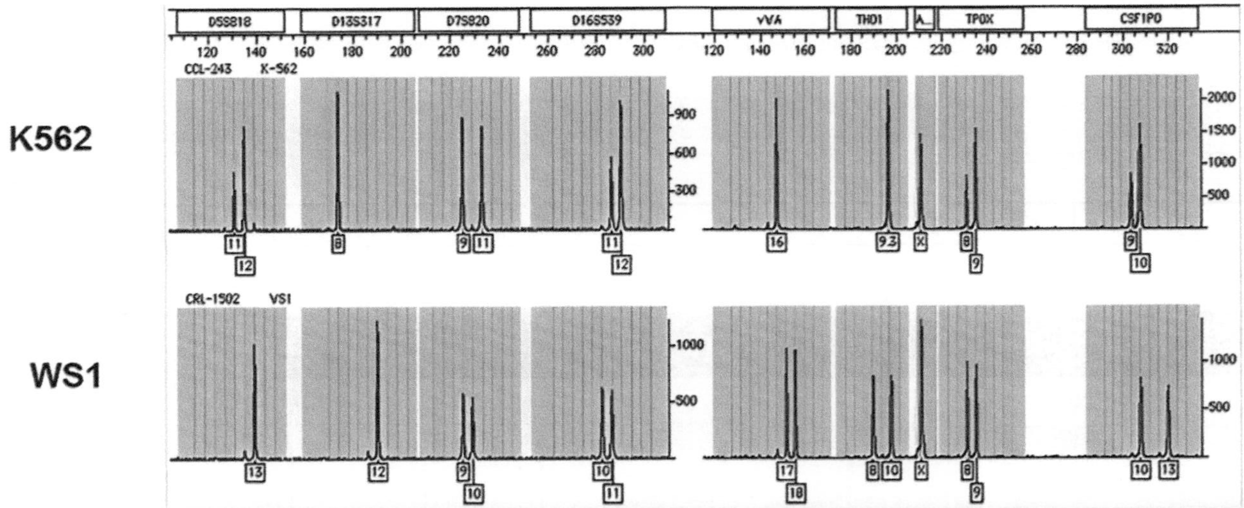

	D5S818	D13S317	D7S820	D16S539	vWA	THO1	Amel.	TPOX	CSF1PO
K562	11, 12	8	9, 11	11, 12	16	9.3	X	8, 9	9, 10
WS1	13	12	9, 10	10, 11	17, 18	8, 10	X	8, 9	10, 13

FIG. 1

Shows an electropherogram of two distinct human cell lines, K562 (chronic myelogenous leukemia) and WS1 (skin fibroblast), which were isolated from two different people. The two cell lines have distinct STR profiles conducted using PowerPlex 1.2 for STR analysis. To generate the database, only numerical values were employed.

Image retrieved from Reid, Y., Storts, D., Riss, T., Minor, L., 2013. Authentication of human cell lines by STR DNA profiling analysis. In: Assay Guidance Manual.

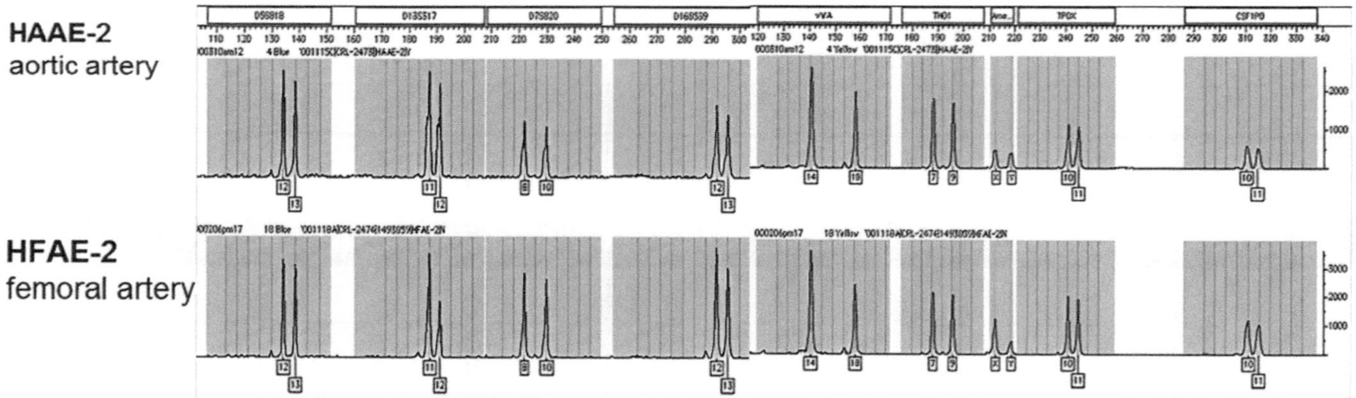

	D5S818	D13S317	D7S820	D16S539	vWA	THO1	Amel.	TPOX	CSF1PO
HAAE-2	12,13	11,12	8,10	12,13	14,18	7,9	X,Y	10,11	10,11
HFAE-2	12,13	11,12	8,10	12,13	14,18	7,9	X,Y	10,11	10,11

FIG. 2

Shows an electropherogram of two related human cell lines; HAAE-2 (human aortic artery) and HFAE-2 (human femoral artery) obtained from the same person. The two cell lines share the same STR characteristics conducted using PowerPlex 1.2 for STR analysis. To generate the database, only numerical values were employed.

Image retrieved from Reid, Y., Storts, D., Riss, T., Minor, L., 2013. Authentication of human cell lines by STR DNA profiling analysis. In: Assay Guidance Manual.

as economical. Primer pairs are designed to hybridize with a single allele or a collection of alleles that are related to it. The amplification of the target sequence is possible under tightly regulated PCR conditions with the help of matching primer pairs.

After PCR, amplified DNA fragments are sorted into different sizes by agarose gel electrophoresis, seen by staining (ethidium bromide or SYBR green) and being exposed to UV light, then recorded using photography for analysis. The presence or absence of particular PCR products in reaction tubes using primers that have known allele specificities is used to evaluate the findings of PCR-SSP. A successful amplification will, in essence, cause a band of a particular length of DNA to emerge on the gel. If there is no amplification, this band is absent. In each PCR, a positive control set of internal primers is utilized. The human growth hormone or β-globin gene has conserved regions that are present in all samples of human DNA. The internal positive control primer pair relates to these regions. The following is a general guide on performing HLA-SSP typing using one of the available commercial kits in the market. Generally, most HLA SSP typing kits have similar protocol with some changes to the volume of reagents for master mix and reaction setup in thermocycler. Thus, the kit manufacturer's instructions must be read carefully for a successful and reliable analysis.

2.2 Protocol

Materials, equipment, and apparatus

1. HLA SSP commercial kit that is, Morgan HLA SSP ABDR Typing Kit (TBG Biotechnology Corp.; New Taipei City, Republic of China) or equivalent
2. DNA samples (concentration of 1–100 ng/μL)
3. *Taq* DNA polymerase (5 U/μL) (PE Applied Biosystems or similar) (Applied Biosystems; Massachusetts, United States)
4. Cresol red dye (Thermo Fisher Scientific; Massachusetts, United States)
5. Thermal cycler with heated cover, i.e., PE 9600, PE9700 Applied Biosystems (Applied Biosystems; Massachusetts, United States)
6. Adjustable pipettes
7. Eppendorf Multipette with combitips (10.0 μL) (Eppendorf; Hamburg, Germany)
8. Filtered pipette tips
9. Distilled water
10. Agarose gel (molecular biology)
11. Gel chamber (suitable for gel with at least 25 pockets)
12. 5× TBE buffer (tris base, boric acid, disodium EDTA)
13. Ethidium bromide solution (10.0 mg/mL)
14. DNA ladder (50–1000 bp marker)
15. Microwave or magnetic stirrer with hotplate
16. UV spectrophotometer (for DNA sample preparation and quality control)
17. Polaroid camera with UV filter and Polaroid film type 667
18. UV transilluminator (approx. 200–300 nm)

2.2.1 Polymerase chain reaction (PCR)

1. Prepare master mix in an Eppendorf reaction tube. The components of reagents across various HLA SSP typing kit are mostly similar, with PCR cocktail (consisting of PCR buffer and primer mix), *Taq* DNA polymerase (5.0 unit/μL), and RNase-free water to make up the reaction mix for each typing test. The 10× primer mix may be dispensed at the time of testing or prealiquoted in the PCR plate depending on the kit. It is suggested to add cresol red dye to the reaction mix to aid with pipetting accuracy and for tracking the sample in gel electrophoresis. Table 3 below depicts the volume of reagents needed per sample.
2. Pipette the reaction mix into the wells of the PCR plate accordingly. Ensure that pipetting is performed accurately and avoid carryover from one well to the other.
3. PCR cover strips are used to secure the PCR strips and avoid evaporation. Start the HLA-SSP program after inserting the tray into the thermal cycler. Table 4 is an example for thermocycler setup for HLA analysis. Please take note that the PCR software below has been modified for Perkin-Elmer thermocyclers. The PE

Table 3 Guide for preparation of reaction mix per sample.

Reagent	Volume (μL)
10× primer mix	1.0
10× PCR buffer	1.0
Taq polymerase (5.0 unit/μL)	0.1
RNase-free water	6.9
DNA sample (10–100 ng/μL)	1.0
Total	**10.0**

Table 4 Thermocycler setup for HLA-SSP.

Program HLA-SSP			
Steps	**Temperature (°C)**	**Time**	**Number of cycles**
1	96.0	2.5 min	1
2	96.0	15 s	10
	65.0°C	60 s	
3	95.0°C	15 s	22
	62.0°C	50 s	
	72.0°C	30 s	
4	4.0	∞ (ready to be removed from instrument)	1

Note: The thermocycler setup is specific according to the kit manufacturer's instructions. If a different HLA SSP typing kit is used, follow the setup suggested by the instructions provided by the manufacturer.

9700 has a temperature precision of ±0.25°C over the range of 35–100°C and a heat/cooling rate for the sample of 1°C/s. It is advised that any additional thermocyclers utilized for this test go through validation for this program. An adapter mat is necessary for thermocyclers without an adjustable pressure plate in order to maximize heat transmission from the heat cover to the PCR tubes.

2.2.2 Gel electrophoresis

1. Prepare the reagents needed:
   a. 5× TBE buffer (0.445 M tris-borate, 0.0125 M EDTA):
      i. Tris(hydroxymethyl)-aminomethane (base)—54.0 g
      ii. Boric acid—27.5 g
      iii. Disodium EDTA—4.65 g add into 1.0 L distilled water, store at room temperature.
   b. Ethidium bromide solution (10.0 mg/mL)
      i. 100.0 mg of ethidium bromide should be dissolved in 10.0 mL of purified water. Store away from light at 2°C to 8°C. Ethidium bromide should be noted as poisonous and mutagenic. When working, put on protective gloves and rinse with water right away if there has been direct contact.
   c. 0.5× TBE buffer
      i. Dilute the 5× TBE buffer solution in a 1:10 ratio with demineralized water.
2. Prepare 2% agarose gel for electrophoresis:
   a. Boil 5.0 g of agarose powder in 250.0 mL of 0.5× TBE until dissolves (boiling using microwave or hot plate with stirrer).
   b. Cool down the solution until <60°C by stirring or short rinse of the flask with tap water. After the solution has cooled, add 4.0 μL of ethidium bromide.
   c. Avoiding bubbles, pour the agarose solution into a gel tray that has been lined and sealed. Keep the combs (≥10.0 μL pockets) for a minimum of 10 min at room temperature after placing them.
   d. Insert the gel into the gel chamber when the agarose has hardened. After removing the combs, apply 0.5× TBE solution over the gel. Make sure the buffer completely encases the gel pockets.
3. Pipette the entire PCR mixtures (approximately 10.0 μL) and a molecular weight standard (50–1000 bp) into the gel pockets. *Note*: ensure loading of PCR mixtures and ladder is done properly and as accurately as possible.
4. Start the electrophoresis process (15–25 min at 8 V/cm). The cresol red (loading dye present in the PCR cocktail) should have a migration distance of 0.7–1.5 cm.

2.2.3 Gel documentation

1. To document and analyze the results, put the gel on a UV transilluminator once the electrophoresis run is over. *Note*: Employ appropriate protection when dealing with UV light.

2.2.4 Results evaluation and interpretation

Depending on the kit, the size range of DNA fragments to be amplified may vary. For Morgan HLA SSP ABDR Typing Kit, the DNA fragments amplified by group- or allele-specific primer pairs were designed to be between 70 and 275 bp in size, with control fragment the size of 600 bp. Thus, the internal control migrates at slower pace compared to the group- or allele-specific band.

Due to the primers' chemical makeup, the HLA characteristics may be positively detected. Whether a certain band is discernible on the gel will affect the interpretation. The size of the amplified DNA fragments should not be considered while evaluating the test, although it may be useful when interpreting the results. After transferring the pattern of the particular bands to the supplied result sheet, the HLA typing result can be read out.

2.2.5 Interpretation of results on the gel based on reaction pattern

The presence of group- or allele-specific band indicates a positive result for the sample tested, while the absence of group- or allele-specific band denotes a negative result for the sample reaction. With the presence of group- and allele-specific band, the internal control band may appear weak or hardly visible. However, since the presence of the specific band is an indication of a successful PCR, the weak or absence of internal control band does not dismiss the outcome of the analysis. Nevertheless, it is important to note that the internal control must always be clearly visible in negative lanes. A test result is invalid if both the internal control band and group- or allele-specific are absence at the end of the analysis as it is an indication of a failed PCR (nonamplification). As always, in the contamination control well (where template DNA was replaced with RNase-free water), the presence of any bands will invalidate the overall test results as it could indicate primer dimer or contamination (Fig. 3).

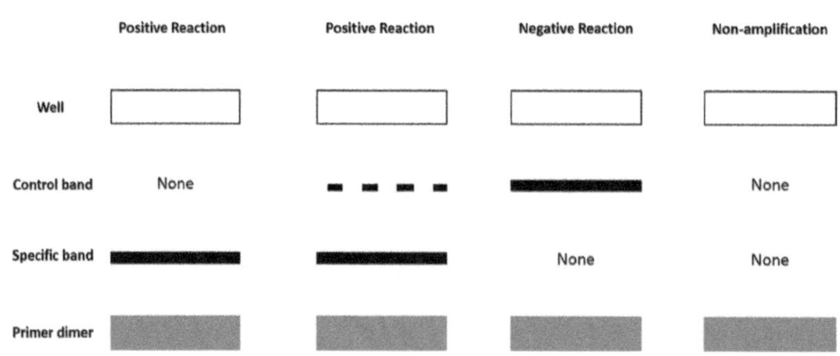

FIG. 3

Interpretation of results based on band appearance on gel.

Image retrieved from the TBG Biotechnology Corp., 2014. Morgan™ HLA ABDR SSP Typing Kit. Retrieved from: http://www.tbgbio.com/en/product/product_detail/2.

3. Karyotyping

3.1 Background and principle

Karyotyping is a method of assessing cellular chromosome morphology and number. It allows us to observe and detect any differences in size, centromere position, and change in banding patterns. It is crucial to find any chromosomal abnormalities because changes to the genome might affect how stem cells express their genes and carry out their biological tasks, which raises the likelihood that the stem cells will become tumorigenic. Therefore, stem cell cultures must be monitored closely for any chromosomal abnormalities, especially for clinical and therapeutic purpose.

3.2 Procedure

The materials needed for this analysis are as follows:

Materials	Equipment and apparatus
1. Cell culture 2. Cell culture growth media (DMEM +10% FBS) (Gibco; Thermo Fisher Scientific Inc., Massachusetts, United States) 3. Colcemid 10.0 μg/mL (Irvine Scientific; California, United States) 4. Hanks' Balanced Salt Solution (HBSS) (Gibco; Thermo Fisher Scientific Inc., Massachusetts, United States) 5. 0.05% Trypsin-EDTA (Gibco; Thermo Fisher Scientific Inc., Massachusetts, United States) 6. 0.5% Trypsin 10× (Santa Cruz Biotechnology; Texas, United States) 7. 0.075 M KCL (Sigma-Aldrich; Darmstadt, Germany) 8. Giemsa stain (Sigma-Aldrich; Darmstadt, Germany) 9. Gurr's buffer (Thermo Fisher Scientific; Massachusetts, United States) 10. Carnoy's fixative (methanol: glacial acetic acid in 3:1 ratio) 11. Distilled water 12. Permount (Fisher Scientific; Thermo Fisher Scientific Inc., Massachusetts, United States) 13. Fetal Bovine Serum (Gibco; Thermo Fisher Scientific Inc., Massachusetts, United States)	1. Incubator 37°C with 5% CO_2 2. Pipette and pipette tips 3. 15-mL conical tube 4. Centrifuge (200×g) 5. Glass slides 6. Coverslip 7. Paper towel 8. Pasteur's pipette 9. Staining rack 10. Light microscope 11. Coplin jars

3.2.1 Chromosome harvesting

1. To the culture flask, add 10.0 μL/mL of colcemid after the cell culture has reached 80% confluency. *Note*: Ensure that the total number of cells is $\geq 2.0 \times 10^6$.
2. For 45 min to 2 h, incubate the cell culture flask at 37°C in an incubator with 5% CO_2.

3. After incubation, harvest cells:
   a. Transfer the spent media into a 15.0-mL centrifuge tube using sterile pipette.
   b. Wash cells with HBSS by gently swirling HBSS on cell surface and remove by pipetting out (discard).
   c. Add 0.05% trypsin-EDTA into the culture flask, ensuring that it covers all surface with cells. Incubate the culture flask at 37°C in a 5% CO_2 incubator for 2–5 min until the cells are detached and dispersed.
   d. Pipette the media collected earlier at step 3a into the culture flask. Perform resuspension of the cells to create a suspension of single cells.
4. Centrifuge the tube with cells at 200×g for 10 min.
5. Aspirate supernatant and leave a small amount (~0.1 mL). Resuspend pellet with the remaining fluid.
6. Add 10.0 mL of prewarmed (37°C) hypotonic solution (0.075 M KCl). Gently pipette up and down to mix cells with hypotonic solution.
7. Incubate cells at 37°C for 10 min.
8. Centrifuge cells at 200×g for 5 min. While centrifuging cells, prepare fixative solution.
   a. Fixative solution: 30.0 mL of methanol and 10.0 mL of glacial acetic acid.
9. Aspirate supernatant; leave a small volume (up to 0.5 mL). Resuspend the cells gently by flicking the tube. Ensure that the pellet is completely resuspended to avoid clumps of cells when adding fixative.
10. Add 7.0 mL of fixative solution:
    a. Vortex the tube while adding the initial 1.0 mL of the fixative drop by drop. It is allowed to add the remaining fixative volume more quickly.
11. For 30 to 60 min, incubate the cells on ice.
12. Replace the fixative twice:
    a. Centrifuge the cells for 5 min at 200×g. Re-suspend cells after removing the supernatant. Add 5.0 mL fixative to the tube.
    b. Centrifuge cells for 5 min at 200×g. Re-suspend cells after removing the supernatant. Add 5.0 mL fixative to the tube. The cells can be kept in 5.0–10.0 mL of fixative at −20°C or utilized for the staining technique (following protocol below).

3.2.2 Slide preparation and solid staining (to observe cells prior to G-banding procedure)

It is advised to perform rapid evaluation on one representative slide for the quality of the harvest before proceeding to G-banding. There are two methods commonly employed to perform quality assessment of the harvest: solid staining and phase contrast microscopy.

Note: Glass slides need to be precleaned by immersing them in ethanol for overnight and cleaned with Kimwipe.

1. Cells should be centrifuged at 200×g for 5 min. Remove most of the supernatant, leaving just 0.3 to 0.5 mL.

2. Resuspend the pellet slowly. Add three drops of the resuspended pellet at a distance of at least 2 in. from the glass slide while tilting them at a 45 degrees angle (allowing the suspension to run across the slide). One drop of fixative should be added on the slide.
3. With a paper towel, the back of the glass slide should be wiped and allowed to dry for at least 10 min. *Note*: The suspension must be allowed to dry fully.
4. Giemsa solution for staining should be prepared (3:1 ratio of Gurr's buffer and Giemsa stain).
5. After the glass slide has dried, place it on a staining rack and use Giemsa staining solution to completely cover the slide's surface. The slide is to be left to stain for 5 min.
6. Use distilled water to clean the slide, then allow it to air-dry.
7. Cover the slide with a coverslip and add four drops of Permount (prevent bubbles in this step). Use paper towels to wipe away excess Permount.
8. Use a light microscope to examine cells at 10× and 100× magnification. The leftover slides can be utilized for additional investigation if there are plenty of evenly distributed metaphase cells.

3.2.3 G-banding (using trypsin and Giemsa)

Note: In areas of DNA with high transcriptional activity, trypsin as a proteolytic enzyme breaks down euchromatic histones. These areas will appear as thin bands following Giemsa staining. Because the bulk of its histones were shielded from trypsinization, extremely compacted chromatin with little to no transcriptional activity would exhibit a black stain after Giemsa staining. It is recommended to perform G-band on one slide first to assess its condition, and then for the remaining slides, modify the time of the trypsin as appropriate and if needed.

1. Set up the following in four separate Coplin jars:
   a. Jar #1: 30.0 mL of 1× HBSS and 4.0 mL of 0.5% Trypsin 10×
   b. Jar #2: 50.0 mL of 1× HBSS
   c. Jar #3: 45.0 mL of 1× HBSS and 5.0 mL of Fetal Bovine Serum (FBS)
   d. Jar #4: 50.0 mL of 1× HBSS
2. Submerse slide in jar #1 for 5 s, then quickly rinse the slide in jar #2.
3. Immerse slide in jar #3 for ≥30 s, then quickly rinse the slide in jar #4. Leave the slide to air-dry.
4. Freshly prepare Giemsa staining solution (Gurr's buffer and Giemsa stain in 3:1 ratio).
5. Set the slide on the staining rack and completely cover it with Giemsa staining solution. Leave to stain for 5 min.
6. Rinse slide in distilled water (if there are multiple slides, rinse the slides in the same order of staining). Let the slide dry for ~10 min. Ensure complete drying of the slide.
7. Cover the slide with a slip and add four drops of Permount (ensure no bubble). Use a light microscope to examine cells at 10× and 100× magnification. Adapt the trypsin timing to the findings on the first slide.

3.2.4 Result interpretation based on microscopic observation

A successful G-banding is indicated by good metaphase spreads (Fig. 4), with characteristic dark and light banding patterns which are important for an accurate chromosomal analysis. To obtain the diagnostic analysis of the chromosomes, the individual chromosomes in the image obtained by G-banding have to be arranged into a standardized format otherwise known as a karyogram (Fig. 5). Briefly, the chromosomes are arranged with their centromeres aligned and numbered in descending order according to their size, with the sex chromosomes placed at the end. The karyogram is then used to identify abnormalities among the chromosomes such as the occurrence of aneuploidy. Further analysis such as Fluorescence In Situ Hybridization (FISH) and Comparative Genomic Hybridization (CGH) can be performed to identify abnormalities in individual gene level.

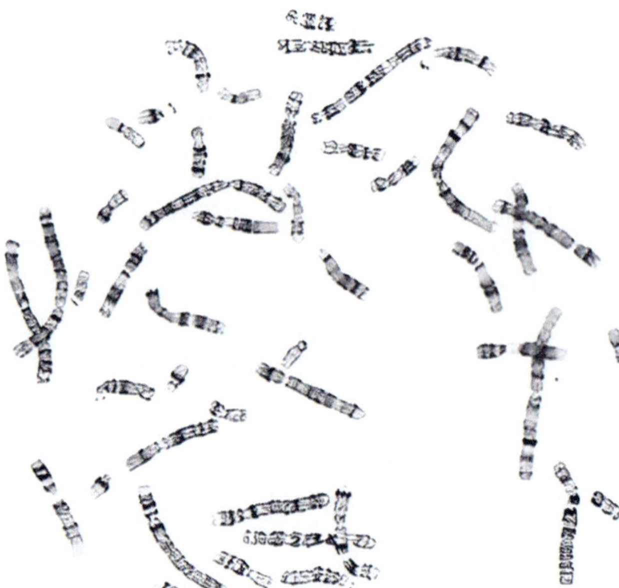

FIG. 4

Example of successful chromosome spread (average length of chromosomes, well spread and easily distinguished from one another).

Image retrieved from Howe, B., Umrigar, A., Tsien, F., 2014. Chromosome preparation from cultured cells. J. Vis. Exp. 83, e50203. doi:10.3791/50203.

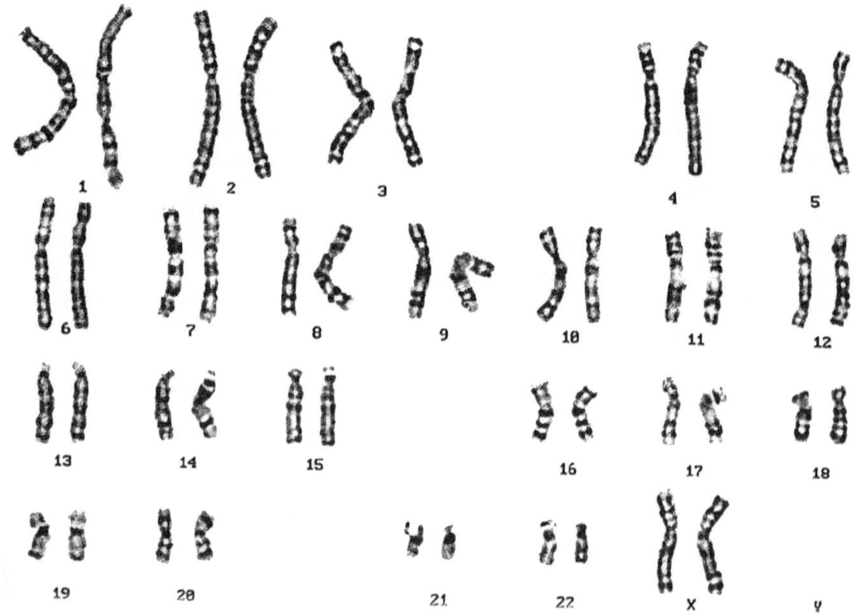

FIG. 5

A karyogram of human female. A karyogram is generated by arranging chromosomal spread into pairs of similar structure and size using their telomeres as a guide. Any abnormalities in the chromosomal structures are observed for further analysis.

Image retrieved from Huang, X.-S., Li, X., Tan, C., Xiao, L., Jiang, H.-O., Zhang, S.-F., Wang, D.-M., Zhang, J.-X., 2010. Genome-wide scanning reveals complex etiology of oculo-auriculo-vertebral spectrum. Tohoku J. Exp. Med. 222 (4), 311–318. doi:10.1620/tjem.222.311. 21150135.

4. Conclusion

This chapter summarized various methods of characterization and has guided with step-by-step procedure in performing the characterization. The selection of characterization method depends on the type of study carried out. It is necessary to select the best applicable characterization mode to establish and confirm on the cell species, purity, and status of the cultures maintained. Characterization plays a major role in identification of transformed cell population in cases such as stored cells. For example, in case of Umbilical Cord Bank, the Cord Blood Cells are stored in liquid nitrogen for numerous years until needed. Whenever there occurs a need for utilizing the stored cells, characterization proves to be an important way to confirm the cells nature, that is, whether the cells are pure without any contamination including the bacterial and transformed cells population.

Further reading

Camper, S.A., Samuelson, L.C., Saunders, T.L., 2021, April 19. ES Cell Chromosome Counting: Laboratory Protocols—U-M Biomedical Research Core Facilities. University of Michigan Medical School Biomedical Research Core Facilities. https://brcf.medicine.umich.edu/cores/transgenic-animal-model/training-education/protocols/chromcounts/.

Chang, Y.H., Wu, K.C., Liu, H.W., Chu, T.Y., Ding, D.C., 2018. Human umbilical cord-derived mesenchymal stem cells reduce monosodium iodoacetate-induced apoptosis in cartilage. Ci Ji Yi Xue Za Zhi 30 (2), 71–80. https://doi.org/10.4103/tcmj.tcmj_23_18.

Chen, G., Yue, A., Ruan, Z., Yin, Y., Wang, R.Z., Ren, Y., Zhu, L., 2014, June 2. Human umbilical cord-derived mesenchymal stem cells do not undergo malignant transformation during long-term culturing in serum-free medium. PLoS ONE. https://journals.plos.org/plosone/article?id=10.1371%2Fjournal.pone.0098565.

https://brcf.medicine.umich.edu/wp-content/uploads/2018/01/Chrom_Prep.pdf.

Index

Note: Page numbers followed by *f* indicate figures and *t* indicate tables.

I

K

L

M

U

W

CPI Antony Rowe
Eastbourne, UK
July 08, 2023